Scott Foresman
SCIENCE

Series Authors

Dr. Timothy Cooney
Professor of Earth Science and
  Science Education
Earth Science Department
University of Northern Iowa
Cedar Falls, Iowa

Michael Anthony DiSpezio
Science Education Specialist
Cape Cod Children's Museum
Falmouth, Massachusetts

Barbara K. Foots
Science Education Consultant
Houston, Texas

Dr. Angie L. Matamoros
Science Curriculum Specialist
Broward County Schools
Ft. Lauderdale, Florida

Kate Boehm Nyquist
Science Writer and Curriculum Specialist
Mount Pleasant, South Carolina

Dr. Karen L. Ostlund
Professor
Science Education Center
The University of Texas at Austin
Austin, Texas

Contributing Authors

Dr. Anna Uhl Chamot
Associate Professor and
  ESL Faculty Advisor
Department of Teacher Preparation
  and Special Education
Graduate School of Education
  and Human Development
The George Washington University
Washington, DC

Dr. Jim Cummins
Professor
Modern Language Centre and
  Curriculum Department
Ontario Institute for Studies in Education
Toronto, Canada

Gale Philips Kahn
Lecturer, Science and Math Education
Elementary Education Department
California State University, Fullerton
Fullerton, California

Vincent Sipkovich
Teacher
Irvine Unified School District
Irvine, California

Steve Weinberg
Science Consultant
Connecticut State
  Department of Education
Hartford, Connecticut

Scott Foresman

Editorial Offices: Glenview, Illinois • Parsippany, New Jersey • New York, New York
Sales Offices: Parsippany, New Jersey • Duluth, Georgia • Glenview, Illinois
Carrollton, Texas • Ontario, California
www.sfscience.com

Content Consultants

Dr. J. Scott Cairns
National Institutes of Health
Bethesda, Maryland

Jackie Cleveland
Elementary Resource Specialist
Mesa Public School District
Mesa, Arizona

Robert L. Kolenda
Science Lead Teacher, K-12
Neshaminy School District
Langhorne, Pennsylvania

David P. Lopath
Teacher
The Consolidated School District
  of New Britain
New Britain, Connecticut

Sammantha Lane Magsino
Science Coordinator
Institute of Geophysics
University of Texas at Austin
Austin, Texas

Kathleen Middleton
Director, Health Education
ToucanEd
Soquel, California

Irwin Slesnick
Professor of Biology
Western Washington University
Bellingham, Washington

Dr. James C. Walters
Professor of Geology
University of Northern Iowa
Cedar Falls, Iowa

Multicultural Consultants

Dr. Shirley Gholston Key
Assistant Professor
University of Houston-Downtown
Houston, Texas

Damon L. Mitchell
Quality Auditor
Louisiana-Pacific Corporation
Conroe, Texas

Classroom Reviewers

Kathleen Avery
Teacher
Kellogg Science/Technology Magnet
Wichita, Kansas

Margaret S. Brown
Teacher
Cedar Grove Primary
Williamston, South Carolina

Deborah Browne
Teacher
Whitesville Elementary School
Moncks Corner, South Carolina

Wendy Capron
Teacher
Corlears School
New York, New York

Jiwon Choi
Teacher
Corlears School
New York, New York

John Cirrincione
Teacher
West Seneca Central Schools
West Seneca, New York

Jacqueline Colander
Teacher
Norfolk Public Schools
Norfolk, Virginia

Dr. Terry Contant
Teacher
Conroe Independent
  School District
The Woodlands, Texas

Susan Crowley-Walsh
Teacher
Meadowbrook Elementary School
Gladstone, Missouri

Charlene K. Dindo
Teacher
Fairhope K-1 Center/Pelican's Nest
  Science Lab
Fairhope, Alabama

Laurie Duffee
Teacher
Barnard Elementary
Tulsa, Oklahoma

Beth Anne Ebler
Teacher
Newark Public Schools
Newark, New Jersey

Karen P. Farrell
Teacher
Rondout Elementary School
  District #72
Lake Forest, Illinois

Anna M. Gaiter
Teacher
Los Angeles Unified School District
  Los Angeles Systemic Initiative
Los Angeles, California

Federica M. Gallegos
Teacher
Highland Park Elementary
Salt Lake School District
Salt Lake City, Utah

Janet E. Gray
Teacher
Anderson Elementary - Conroe ISD
Conroe, Texas

Karen Guinn
Teacher
Ehrhardt Elementary School - KISD
Spring, Texas

Denis John Hagerty
Teacher
Al Ittihad Private Schools
Dubai, United Arab Emirates

Judith Halpern
Teacher
Bannockburn School
Deerfield, Illinois

Debra D. Harper
Teacher
Community School District 9
Bronx, New York

Gretchen Harr
Teacher
Denver Public Schools - Doull School
Denver, Colorado

Bonnie L. Hawthorne
Teacher
Jim Darcy School
School Dist #1
Helena, Montana

Marselle Heywood-Julian
Teacher
Community School District 6
New York, New York

Scott Klene
Teacher
Bannockburn School 106
Bannockburn, Illinois

Thomas Kranz
Teacher
Livonia Primary School
Livonia, New York

Tom Leahy
Teacher
Coos Bay School District
Coos Bay, Oregon

Mary Littig
Teacher
Kellogg Science/Technology Magnet
Wichita, Kansas

Patricia Marin
Teacher
Corlears School
New York, New York

Susan Maki
Teacher
Cotton Creek CUSD 118
Island Lake, Illinois

Efraín Meléndez
Teacher
East LA Mathematics Science
  Center LAUSD
Los Angeles, California

Becky Mojalid
Teacher
Manarat Jeddah Girls' School
Jeddah, Saudi Arabia

Susan Nations
Teacher
Sulphur Springs Elementary
Tampa, Florida

Brooke Palmer
Teacher
Whitesville Elementary
Moncks Corner, South Carolina

Jayne Pedersen
Teacher
Laura B. Sprague
  School District 103
Lincolnshire, Illinois

Shirley Pfingston
Teacher
Orland School Dist 135
Orland Park, Illinois

Teresa Gayle Rountree
Teacher
Box Elder School District
Brigham City, Utah

Helen C. Smith
Teacher
Schultz Elementary
Klein Independent School District
Tomball, Texas

Denette Smith-Gibson
Teacher
Mitchell Intermediate, CISD
The Woodlands, Texas

Mary Jean Syrek
Teacher
Dr. Charles R. Drew Science
  Magnet
Buffalo, New York

Rosemary Troxel
Teacher
Libertyville School District 70
Libertyville, Illinois

Susan D. Vani
Teacher
Laura B. Sprague School
School District 103
Lincolnshire, Illinois

Debra Worman
Teacher
Bryant Elementary
Tulsa, Oklahoma

Dr. Gayla Wright
Teacher
Edmond Public School
Edmond, Oklahoma

Activity and Safety Consultants

Laura Adams
Teacher
Holley-Navarre Intermediate
Navarre, Florida

Dr. Charlie Ashman
Teacher
Carl Sandburg Middle School
Mundelein District #75
Mundelein, Illinois

Christopher Atlee
Teacher
Horace Mann Elementary
Wichita Public Schools
Wichita, Kansas

David Bachman
Consultant
Chicago, Illinois

Sherry Baldwin
Teacher
Shady Brook
Bedford ISD
Euless, Texas

Pam Bazis
Teacher
Richardson ISD
Classical Magnet School
Richardson, Texas

Angela Boese
Teacher
McCollom Elementary
Wichita Public Schools USD #259
Wichita, Kansas

Jan Buckelew
Teacher
Taylor Ranch Elementary
Venice, Florida

Shonie Castaneda
Teacher
Carman Elementary, PSJA
Pharr, Texas

Donna Coffey
Teacher
Melrose Elementary - Pinellas
St. Petersburg, Florida

Diamantina Contreras
Teacher
J.T. Brackenridge Elementary
San Antonio ISD
San Antonio, Texas

Susanna Curtis
Teacher
Lake Bluff Middle School
Lake Bluff, Illinois

Karen Farrell
Teacher
Rondout Elementary School,
Dist. #72
Lake Forest, Illinois

Paul Gannon
Teacher
El Paso ISD
El Paso, Texas

Nancy Garman
Teacher
Jefferson Elementary School
Charleston, Illinois

Susan Graves
Teacher
Beech Elementary
Wichita Public Schools USD #259
Wichita, Kansas

Jo Anna Harrison
Teacher
Cornelius Elementary
Houston ISD
Houston, Texas

Monica Hartman
Teacher
Richard Elementary
Detroit Public Schools
Detroit, Michigan

Kelly Howard
Teacher
Sarasota, Florida

Kelly Kimborough
Teacher
Richardson ISD
Classical Magnet School
Richardson, Texas

Mary Leveron
Teacher
Velasco Elementary
Brazosport ISD
Freeport, Texas

Becky McClendon
Teacher
A.P. Beutel Elementary
Brazosport ISD
Freeport, Texas

Suzanne Milstead
Teacher
Liestman Elementary
Alief ISD
Houston, Texas

Debbie Oliver
Teacher
School Board of Broward County
Ft. Lauderdale, Florida

Sharon Pearthree
Teacher
School Board of Broward County
Ft. Lauderdale, Florida

Jayne Pedersen
Teacher
Laura B. Sprague School
District 103
Lincolnshire, Illinois

Sharon Pedroja
Teacher
Riverside Cultural
Arts/History Magnet
Wichita Public Schools USD #259
Wichita, Kansas

Marcia Percell
Teacher
Pharr, San Juan, Alamo ISD
Pharr, Texas

Shirley Pfingston
Teacher
Orland School Dist #135
Orland Park, Illinois

Sharon S. Placko
Teacher
District 26, Mt. Prospect
Mt. Prospect, IL

Glenda Rall
Teacher
Seltzer Elementary
USD #259
Wichita, Kansas

Nelda Requenez
Teacher
Canterbury Elementary
Edinburg, Texas

Dr. Beth Rice
Teacher
Loxahatchee Groves
Elementary School
Loxahatchee, Florida

Martha Salom Romero
Teacher
El Paso ISD
El Paso, Texas

Paula Sanders
Teacher
Welleby Elementary School
Sunrise, Florida

Lynn Setchell
Teacher
Sigsbee Elementary School
Key West, Florida

Rhonda Shook
Teacher
Mueller Elementary
Wichita Public Schools USD #259
Wichita, Kansas

Anna Marie Smith
Teacher
Orland School Dist. #135
Orland Park, Illinois

Nancy Ann Varneke
Teacher
Seltzer Elementary
Wichita Public Schools USD #259
Wichita, Kansas

Aimee Walsh
Teacher
Rolling Meadows, Illinois

Ilene Wagner
Teacher
O.A. Thorp Scholastic Acacemy
Chicago Public Schools
Chicago, Illinois

Brian Warren
Teacher
Riley Community Consolidated
School District 18
Marengo, Illinois

Tammie White
Teacher
Holley-Navarre
Intermediate School
Navarre, Florida

Dr. Mychael Willon
Principal
Horace Mann Elementary
Wichita Public Schools
Wichita, Kansas

Inclusion Consultants

Dr. Eric J. Pyle, Ph.D.
Assistant Professor, Science Education
Department of Educational Theory
and Practice
West Virginia University
Morgantown, West Virginia

Dr. Gretchen Butera, Ph.D.
Associate Professor, Special Education
Department of Education Theory
and Practice
West Virginia University
Morgantown, West Virginia

Bilingual Consultant

Irma Gomez-Torres
Dalindo Elementary
Austin ISD
Austin, Texas

Bilingual Reviewers

Mary E. Morales
E.A. Jones Elementary
Fort Bend ISD
Missouri City, Texas

Gabriela T. Nolasco
Pebble Hills Elementary
Ysleta ISD
El Paso, Texas

Maribel B. Tanguma
Reed and Mock Elementary
San Juan, Texas

Yesenia Garza
Reed and Mock Elementary
San Juan, Texas

Teri Gallegos
St. Andrew's School
Austin, Texas

iii

Unit B
Physical Science

Unit C
Earth Science

Unit D
Human Body

Using Scientific Methods for Science Inquiry

Scientists try to solve many problems. Scientists study problems in different ways, but they all use scientific methods to guide their work. Scientific methods are organized ways of finding answers and solving problems. Scientific methods include the steps shown on these pages. The order of the steps or the number of steps used may change. You can use these steps to organize your own scientific inquiries.

Does the amount of water in a bottle affect the pitch of the sound the bottle makes?

State the Problem
The problem is the question you want to answer. Curiosity and inquiry have resulted in many scientific discoveries. State your problem in the form of a question.

Formulate Your Hypothesis
Your hypothesis is a possible answer to your problem. Make sure your hypothesis can be tested. Your hypothesis should take the form of a statement.

◄ A bottle with a lot of water will make a sound with a low pitch.

Identify and Control the Variables
For a fair test, you must select which variable to change and which variables to control. Choose one variable to change when you test your hypothesis. Control the other variables so they do not change.

▲ Use four bottles of the same size. Leave the first bottle empty. Put a little water in the second bottle. Fill the third bottle halfway. Fill the fourth almost full with water.

Test Your Hypothesis

Do experiments to test your hypothesis. You may need to repeat experiments to make sure your results remain consistent. Sometimes you conduct a scientific survey to test a hypothesis.

Tap each bottle gently with a spoon. ▼

Collect Your Data

As you test your hypothesis, you will collect data about the problem you want to solve. You may need to record measurements. You might make drawings or diagrams. Or you may write lists or descriptions. Collect as much data as you can while testing your hypothesis.

Lower   High   Lowest   Low

Interpret Your Data

By organizing your data into charts, tables, diagrams, and graphs, you may see patterns in the data. Then you can decide what the information from your data means.

Amount of water in bottle	Pitch
Empty	Highest
A little water	Lower
Half full	Still lower
Full	Lowest

State Your Conclusion

Your conclusion is a decision you make based on evidence. Compare your results with your hypothesis. Based on whether or not your data supports your hypothesis, decide if your hypothesis is correct or incorrect. Then communicate your conclusion by stating or presenting your decision.

Bottles with more water make a sound with a lower pitch.

Inquire Further

Use what you learn to solve other problems or to answer other questions that you might have. You may decide to repeat your experiment, or to change it based on what you learned.

▲ *Does the size of the bottle make a difference in the pitch of a sound?*

Using Process Skills for Science Inquiry

These 12 process skills are used by scientists when they do their research. You also use many of these skills every day. For example, when you think of a statement that you can test, you are using process skills. When you gather data to make a chart or graph, you are using process skills. As you do the activities in your book, you will use these same process skills.

Observing
Use one or more of your senses—seeing, hearing, smelling, touching, or tasting—to gather information about objects or events.

I see..., I smell..., I hear..., It feels like..., I never taste without permission!

Communicating
Share information about what you learn using words, pictures, charts, graphs, and diagrams.

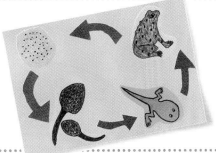

Classifying
Arrange or group objects according to their common properties.

◀ *Living things in Group 1.*

Nonliving things in Group 2. ▶

Estimating and Measuring
Make an estimate about an object's properties, then measure and describe the object in units.

Inferring
Draw a conclusion or make a reasonable guess based on what you observe, or from your past experiences.

The juice was cold this morning... It's still cold... The bottle is an...

Predicting

Form an idea about what will happen based on evidence.

◀ *Predict what will happen after 15 minutes.*

Making Operational Definitions

Define or describe an object or event based on your experiences with it.

A simple machine has no parts that move and it helps... ▶

Making and Using Models

Make real or mental representations to explain ideas, objects, or events.

◀ *It's different from a real food chain because... The model is like a real food chain because...*

Formulating Questions and Hypotheses

Think of a statement that you can test to solve a problem or to answer a question about how something works.

If you put one bean seed in loam and another one in sand, the plant will grow better in... ▶

Collecting and Interpreting Data

Gather observations and measurements into graphs, tables, charts, or diagrams. Then use the information to solve problems or answer questions.

The seed grew better in loam.

Identifying and Controlling Variables

Change one factor that may affect the outcome of an event while holding other factors constant.

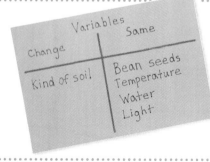

Experimenting

Design an investigation to test a hypothesis or to solve a problem. Then form a conclusion.

I'll write a clear procedure so that other students could repeat the experiment.

⍰ Science Inquiry

Throughout your science book, you will ask questions, do investigations, answer your questions, and tell others what you have learned. Use the descriptions below to help you during your scientific inquiry.

> Does soup powder dissolve faster in hot water or cold water?

❶ Ask a question about objects, organisms, and events in the environment.

You will find the answer to your question from your own observations and investigations and from reliable sources of scientific information.

❷ Plan and conduct a simple investigation.

The kind of investigation you do depends on the question you ask. Kinds of investigations include describing objects, events, and organisms; classifying them; and doing a fair test or experiment.

❸ Use simple equipment and tools to gather data and extend the senses.

Equipment and tools you might use include rulers and meter sticks, compasses, thermometers, watches, balances, spring scales, hand lenses, microscopes, cameras, calculators, and computers.

❹ Use data to construct a reasonable explanation.

Use the information that you have gathered to answer your question and support your answer. Compare your answer to scientific knowledge, your experiences, and the observations of others.

❺ Communicate investigations and explanations.

Share your work with others by writing, drawing, or talking. Describe your work in a way that others could repeat your investigation.

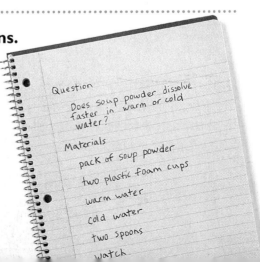

Question
Does soup powder dissolve faster in warm or cold water?

Materials
pack of soup powder
two plastic foam cups
warm water
cold water
two spoons
watch

Unit A
Life Science

A 1

Science and Technology

In Your World!

No Soil Needed!

What? No soil? These plants do not need any soil. They grow in a hydroponic, or "water working," garden. Plants that grow this way have their roots in water mixed with plant food. This allows the plant parts to grow bigger, faster. You'll learn more about plant parts and how plants grow in **Chapter 1 How Plants Live and Grow.**

Pip! Pip! I'm Hatching Now!

Cheep! This incubator is almost like a mother hen. It has a heater to keep chick eggs warm before and after they hatch. Special pumps move fresh, moist air around the eggs and young chicks. You'll learn more about animals and how they start out as eggs in **Chapter 2 How Animals Grow and Change.**

Night Is Day and Day Is Night!

Want to see bats? You might have a problem. They are active only at night. However, some zoos build habitats that are dark during the day. Through dim lights, you may see the bats moving. At night, the habitats are lit to "fool" the bats into sleeping. You'll read more about habitats in **Chapter 3 Living Things and Their Environments.**

This "Living Machine" Eats Pollution!

Pass the snails, please! Dr. John Todd uses plants, snails, clams, algae, and fish to clean polluted water. His Living Machines™ have tanks, pools, and "marshes." Inside them, algae, bacteria, plants, and sunlight change water pollutants into harmless substances. You'll learn more about ways to protect habitats in **Chapter 4 Changing Environments.**

Those Amazing Plants!

Food, wood, and many fabrics come from plants. Plants are more than just useful living things. They make the world more colorful, and they can smell good too!

Chapter 1
How Plants Live and Grow

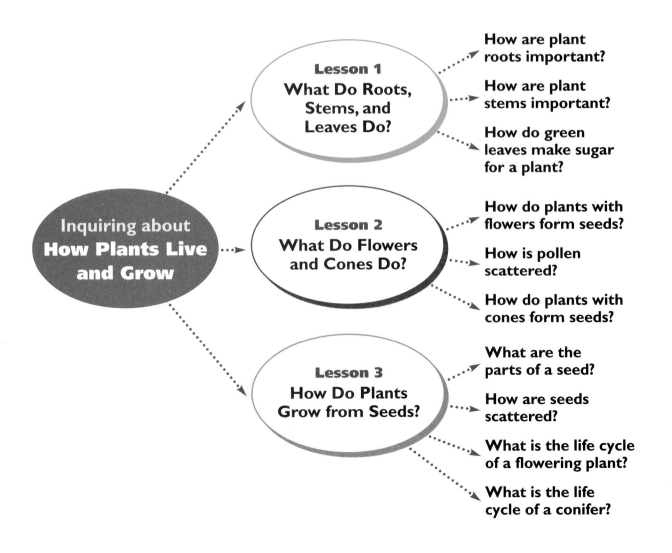

Inquiring about **How Plants Live and Grow**

Lesson 1
What Do Roots, Stems, and Leaves Do?

How are plant roots important?

How are plant stems important?

How do green leaves make sugar for a plant?

Lesson 2
What Do Flowers and Cones Do?

How do plants with flowers form seeds?

How is pollen scattered?

How do plants with cones form seeds?

Lesson 3
How Do Plants Grow from Seeds?

What are the parts of a seed?

How are seeds scattered?

What is the life cycle of a flowering plant?

What is the life cycle of a conifer?

Copy the chapter graphic organizer onto your own paper. This organizer shows you what the whole chapter is all about. As you read the lessons and do the activities, look for answers to the questions and write them on your organizer.

Exploring Parts of a Plant

Process Skills

- observing
- communicating

Materials

- plant
- newspaper
- hand lens

Explore

① **Observe** the stem of the plant. Is it smooth or rough? long or short? Draw and write what you observe.

② Look at the leaves. How many are there? Do they feel thick or thin? Are they shiny or dull? Draw and write what you observe.

③ Hold the plant over a piece of newspaper. Hold the stem between your fingers as shown, and turn the pot over. Gently tap the bottom of the pot and carefully pull the plant out.

④ Place the plant on the newspaper. Remove the soil. Observe the roots with the hand lens. Draw and write what you observe. Repot the plant.

Reflect

Communicate. Compare and contrast your observations with those of other groups.

? Inquire Further

What do the leaves and stems of plants in your neighborhood look like? Develop a plan to answer this or other questions you may have.

Sequencing

Perhaps you have noticed that many things happen in a certain order, or sequence. Placing things in the order in which they happen is called **sequencing**. In the following chapter, *How Plants Live and Grow*, you will learn about the steps in the process of plant growth. These steps happen in a sequence. As you read Lesson 1, *What Do Roots, Stems, and Leaves Do?*, ask yourself what sequence of steps green plants go through to make sugar.

Reading Vocabulary

sequence (sē′kwəns), one thing happening after another

Example

Suppose you wanted to make a peanut butter and jelly sandwich. First, you get two pieces of bread. Second, you spread peanut butter on one piece of bread. Third, you spread jelly on the other piece. Fourth, you put the two pieces of bread together. You have followed a sequence of steps to make your sandwich.

Talk About It!

1. The pictures below show how a seed grows into a plant. They are not in the correct order. On a piece of paper, put the pictures in the correct sequence. Write the letter of the picture that goes first, second, third, and fourth.

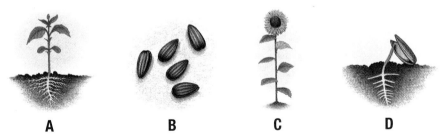

A   B   C   D

2. Make a list of five things you do every day. Number the things on your list in the sequence in which you do them.

You will learn:

- how roots are important to plants.
- how stems are important to plants.
- how green leaves make sugar for a plant.

Glossary

mineral (min′ər əl), a natural, nonliving material that can be found in soil

Lesson 1

What Do Roots, Stems, and Leaves Do?

WOW! It was difficult to pull that weed from the ground! Now you can see why. Look at how long the root is! Why do plants need roots anyway?

Plant Roots

Look at the plant roots in the picture. They may look different, but roots are important to plants in the same ways.

Plant roots grow down into the soil. They hold plants tightly in the ground. Roots take water and minerals from the soil. **Minerals** are natural, nonliving materials that can be found in soil. Plants need water and minerals to live and grow.

Dandelions have one large root that grows straight down in the soil, with thin roots growing out from it.

The thin roots of grass grow out in many directions. These roots may grow to be as long as the large dandelion root.

Plant Stems

Most plants have stems. Stems hold up the leaves, flowers, and fruit of a plant. Most stems, such as the thick, woody stem of the tree in the picture, grow straight up from the ground. Some stems, such as the thin stem of the vine, grow along the ground. Vines can also grow up and over objects.

You know that plant roots take water and minerals from the soil. Plant stems have tiny tubes that carry the water and minerals from the roots to the rest of the plant. Stems also carry sugars made in plant leaves to other parts of the plant.

The thick stem of a tree and the thin stem of a vine ▶

Glossary

carbon dioxide
(kär′bən dī ok′sīd),
a gas in the air that
plants use to make
food

Plant Leaves

Think of the different plant leaves you have seen. Some are wide and flat. Others are long and slender. Still other leaves look like needles. Green plants use energy from the sun to make food in their leaves. The food the leaves make is a kind of sugar. The picture shows the way a plant makes sugar.

Gas from the Air
Carbon dioxide, a gas in the air, goes into tiny openings in the leaves.

Water from the Soil
The roots of a plant take in water from the soil. The water goes through the roots and stem to the leaves of the plant.

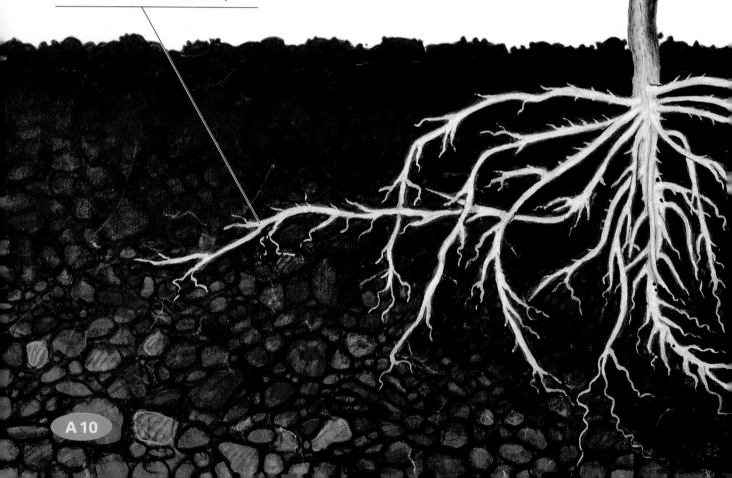

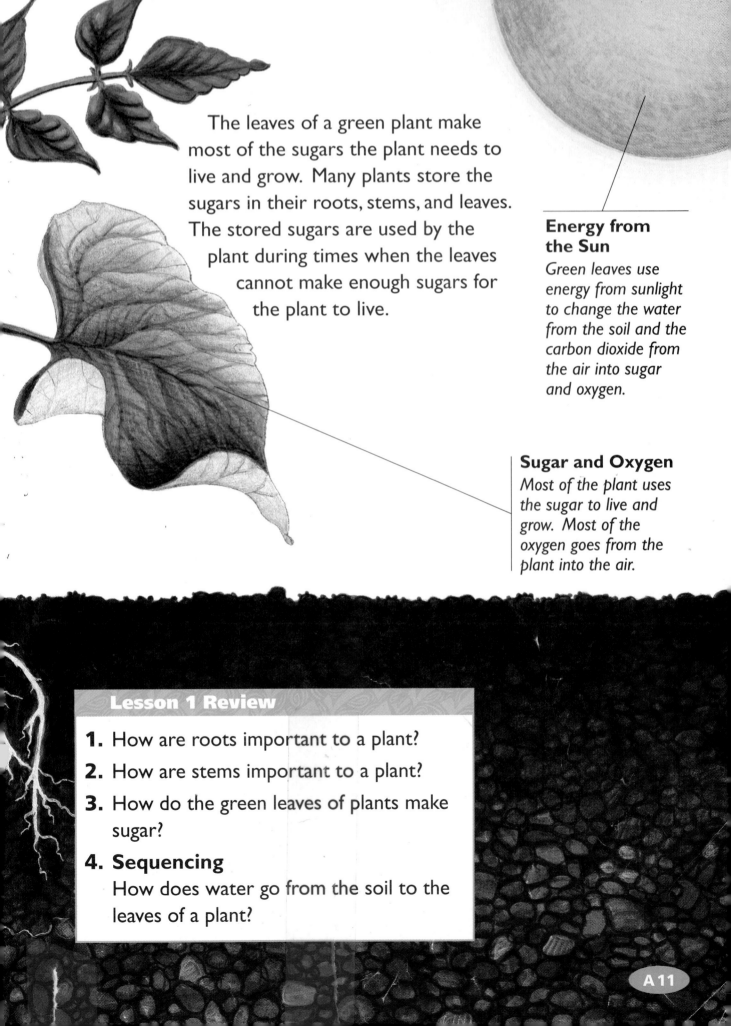

The leaves of a green plant make most of the sugars the plant needs to live and grow. Many plants store the sugars in their roots, stems, and leaves. The stored sugars are used by the plant during times when the leaves cannot make enough sugars for the plant to live.

Energy from the Sun

Green leaves use energy from sunlight to change the water from the soil and the carbon dioxide from the air into sugar and oxygen.

Sugar and Oxygen

Most of the plant uses the sugar to live and grow. Most of the oxygen goes from the plant into the air.

Lesson 1 Review

1. How are roots important to a plant?

2. How are stems important to a plant?

3. How do the green leaves of plants make sugar?

4. Sequencing
How does water go from the soil to the leaves of a plant?

Observing Fruits and Seeds

Process Skills

Process Skills

- observing
- inferring

Materials

- a variety of whole and cut fruits
- paper towel
- metal spoon
- hand lens
- colored pencils or crayons

Getting Ready

In this activity you can find out how seeds and fruits from different plants are alike and different.

Follow This Procedure

1 Make a chart like the one shown. Use your chart to record your observations.

Name of fruit			
Observations and drawing of fruit			
Number of seeds			
Observations and drawing of seeds			

2 Observe a whole piece of fruit. Record your observations and draw a picture of the fruit.

3 Place a cut piece of fruit on a paper towel (Photo A). Use the spoon to remove the seeds. Count the number of seeds in one piece. Use the number to estimate the total number of seeds in the whole fruit. Record the number.

Photo A

④ Use your hand lens to observe one seed from the fruit (Photo B). What shape is it? What color is it? Write a description and draw a picture of the seed.

 Safety Note *Do not eat the fruit. Keep the seeds away from your nose and mouth.*

⑤ Repeat steps 2–4 until you have examined at least three different fruits and their seeds.

Self-Monitoring
Have I written a description and made drawings for each fruit I observed?

Interpret Your Results

1. Compare the number of seeds found in each fruit. Which fruit had the most seeds? Which fruit had the least seeds?

Photo B

2. Compare and contrast the seeds. How are they alike? How are they different?

3. Every fruit you observed had seeds. Make an **inference** about the relationship between fruits and seeds.

❓ Inquire Further

How can you find out if the seeds you observed will grow? Develop a plan to answer this or other questions you may have.

Self-Assessment

- I followed instructions to **observe** fruits and seeds.
- I counted the seeds in each piece of fruit and estimated the number of seeds in the whole fruit.
- I recorded and made drawings of my observations.
- I compared and contrasted the different seeds.
- I made an **inference** about the relationship between fruits and seeds.

You will learn:

- how flowers form seeds.
- how pollen is scattered.
- how cones form seeds.

Glossary

petal (pet′l), an outside part of a flower that is often colored

pollen (pol′ən), a fine, yellowish powder in a flower

▲ *The girl has flowers with brightly colored petals. Pollen stuck to her nose when she smelled the flowers.*

Lesson 2

What Do Flowers and Cones Do?

Smell those beautiful flowers! Many of the flowers in your garden grew from seeds. Trees and grass grow from seeds too. So what do flowers and cones have to do with seeds?

How Flowers Form Seeds

Look at the different flowers below. Seeds grow inside flowers. The picture shows that a **petal** is an outside part of a flower. Seeds are formed inside the petals at the center of the flower. Notice where **pollen,** a fine, yellowish powder, is made. Pollen must first be moved to the tip of the stemlike part, then to the center of the flower, before seeds can begin to form.

Parts of a flower ▼

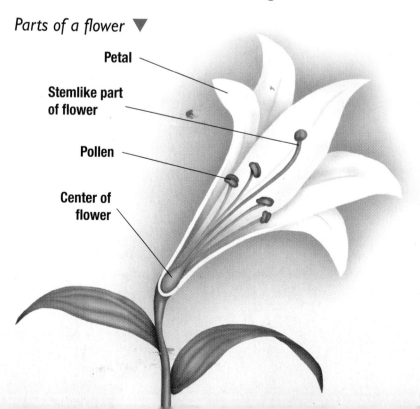

Petal

Stemlike part of flower

Pollen

Center of flower

Suppose pollen had been moved to the center of the flower in the picture on the right. Soon seeds begin to grow. As the seeds grow, the center of the flower swells to form a fruit around the seeds. A fruit is the part of the plant where the seeds are. Apples, peaches, peppers, and tomatoes are examples of fruits. The fruit protects the seeds inside.

Look at the picture of the fruit of a pumpkin plant. As the fruit grows, the flower petals dry up and fall off of the plant. Most fruits are formed this way. Notice the pumpkin seeds in the picture below. The pumpkin that formed around the seeds protects them. You have probably seen seeds inside different fruits. The seeds that form in a fruit can be planted. Some of them will grow to make new plants.

▲ Pollen has been moved to the center of the flower on the pumpkin plant. Seeds will begin to grow.

▲ The fruit of the pumpkin is growing around the seeds.

Seeds of a pumpkin ▼

Scattering Pollen

The color and the smell of a flower attract insects and birds to the flower. Insects and birds can carry pollen from flower to flower. The bee in the picture has landed on a flower and brushed against its pollen. Pollen has a sticky coating. It sticks to the bee's body. Then the bee flies to another flower. Some of the pollen falls off of the bee. The pollen sticks to the second flower. The bee may move the pollen to the stemlike part of the second flower. If this happens, the bee **pollinates** the flower.

Pollen can also be scattered by wind. Wind blows pollen off of a flower. The pollen travels through the air. If the pollen lands on another flower, the flower is pollinated. The pollen from the ragweed in the picture is scattered by wind. Many grasses and trees are pollinated by wind. Wind also pollinates corn.

Bees help pollinate flowers. Suppose this bee flies to another flower of the same kind. Pollen from the bee's body might stick to the flower. Seeds might form. ▼

▲ *Pollen from ragweed is scattered by the wind.*

A 16

How Cones Form Seeds

You have learned how flowers form seeds. Some cones also form seeds. Other cones produce pollen. Pollen from cones is scattered by wind. Wind blows the pollen from the cones into the air. Some of the pollen lands on other cones and pollinates them. Seeds begin to grow inside the pollinated cones. In dry weather, the scales of a cone open. The seeds fall from the cone. Notice the seeds falling from the cone in the picture.

Most trees with cones keep their leaves all year. These trees are called evergreens. The small leaves of evergreens look like needles. Pine and fir trees are examples of evergreens.

Lesson 2 Review

1. How do plants with flowers form seeds?

2. What are two ways pollen can be scattered?

3. How do plants with cones form seeds?

4. **Sequencing**
   Suppose a bee pollinates a flower. Describe how a fruit might form.

▲ The winged seeds inside the cone twirl like a helicopter to the ground.

Investigating Light and Plant Growth

Process Skills

- observing
- inferring
- communicating

Materials

- masking tape
- marker
- 2 resealable plastic bags
- seed starter mix
- large spoon
- radish seeds
- paper towel
- hand lens

Getting Ready

In this activity you can find out how light affects plant growth.

Seeds should be planted just below the surface of the soil.

Follow This Procedure

1 Make a chart like the one shown. The chart should start on Day 1 and end on Day 10. Use your chart to record your observations.

Observations

Day	Plants in light (bag A)	Plants in dark (bag B)
1 (after germination)		
2		
3		

2 Use masking tape and a marker to label one plastic bag A and the other plastic bag B.

3 Use the spoon to fill each bag half full with moistened seed starter mix. Place four radish seeds in each bag. Cover the seeds with a thin layer of the mix (Photo A). Seal the bags.

⚠ **Safety Note** Do not put the seeds in your mouth. Wash your hands after you have planted the seeds.

4 Place bag A in a well-lighted place. Place bag B in a dark place.

5 **Observe** the bags until some of the seeds germinate. Draw a picture of what you see. Describe what you see.

Photo A

6 Repeat step 5 for nine more days. Leave the bags opened if the plants grow to the top of the bags. Keep the seed starter mix moist.

7 Place a piece of paper towel on your desk. Remove the plants from bag A. Place them on the paper towel. Use a hand lens to observe the roots of each plant (Photo B). Draw a picture of one plant and describe what you see. Repeat for bag B.

Interpret Your Results

1. How are the plants grown in the light different from the plants grown in the dark?

2. Make an **inference.** How is a plant affected by light?

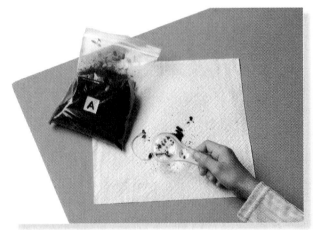

Photo B

3. Communicate. Compare and contrast your observations with those of your classmates.

 **Inquire Further**

What will happen if you keep one plant in the dark and one plant in the light for several weeks? Develop a plan to answer this or other questions you may have.

Self-Assessment

- I followed instructions to test how light affects plant growth.
- I **observed** plants grown in the dark and in the light.
- I recorded my observations.
- I made an **inference** about how light affects plant growth.
- I **communicated** by comparing and contrasting my observations with those of other students.

You will learn:

- the parts of a seed.
- how seeds are scattered.
- the life cycle of a flowering plant.
- the life cycle of a conifer.

Glossary

seed coat, the outside covering of a seed

seed leaf, the part inside each seed that contains stored food

How Do Plants Grow from Seeds?

A coconut is a seed! Oh, come on! It's so big! Think of a grass seed. It's so small! Think of other seeds. Some are oval. Others are pointed. Some are speckled. Others are solid green. Seeds are so different—or are they?

Seed Parts

Seeds come in many sizes, shapes, colors, and patterns, but they are alike in one way. Seeds can grow into new plants.

The thin, hard outer covering of a seed is called the **seed coat.** The seed coat protects the seed. Notice the developing plant in the drawing of the bean seed below. A tiny new plant is inside every seed. When the developing plant starts to grow, it uses food stored in the **seed leaf.**

Parts of a seed ▼

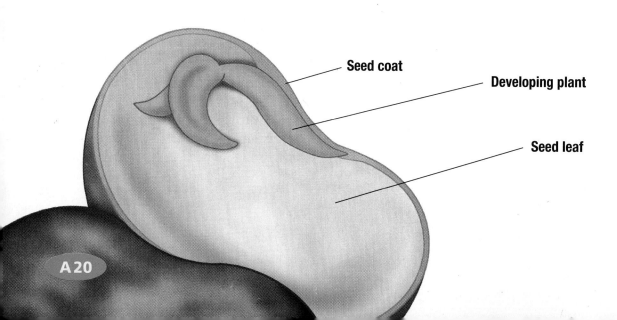

Seed coat

Developing plant

Seed leaf

Scattering Seeds

Some seeds fall to the ground around the parent plant. Other seeds are scattered. Scattering seeds helps each plant get enough space and sunlight to grow. Seeds can be scattered many different ways.

Some seeds are scattered by wind. Dandelion and cattail seeds have light, feathery parts that drift easily in the wind. Wind also helps scatter the maple tree seed in the picture.

Water can scatter some kinds of seeds. Water lily seeds float on lakes and streams. Each coconut fruit in the picture has a seed inside. Coconuts can float long distances in the sea.

Animals can also scatter seeds. Mice and squirrels carry seeds away and bury them in the ground. Some seeds stick to the fur or feathers of animals. Animals can carry seeds in the mud that sticks to their claws or feet. The seeds fall off the animal as it moves.

The winglike part of the maple seed helps it travel through the air. ▼

◀ *Coconuts are light enough to float. Sometimes a coconut fruit can float for miles in the water before it washes up on a warm beach. Then the seed inside might sprout and grow into a tree.*

Glossary

life cycle (sī′kəl), the stages in the life of a living thing

germinate (jėr′mə nāt), to begin to grow and develop

seedling (sēd′ling), a young plant that grows from a seed

The Life Cycle of a Flowering Plant

All living things have a life cycle. A **life cycle** is all the stages in the life of a living thing. Even you have a life cycle. Think about how you looked in your baby pictures. Think about how you look now. As time passes, you go through different stages. The drawings on these two pages show the life cycle of a flowering plant.

Seeds

Seeds grow in the center of the pollinated flower. The petals of the flower dry up and fall off. Seeds fall to the ground. Some seeds are scattered. The seeds are the beginning of a new life cycle.

Germination

A seed **germinates** when the tiny plant inside begins to grow and develop. A seed needs the proper temperature, enough air, and the right amount of water to germinate. The seed soaks up the water and swells until the seed coat splits. A root grows out of the split seed coat down into the soil. Then the stem begins to grow upward.

Seedlings

A **seedling,** or young plant, begins to grow out of the ground.

Growth and Pollination

The plant is fully grown. The new plant looks like the parent plant. The flower on the plant grows and opens. Birds and insects carry pollen from flower to flower. The flower is pollinated.

A23

The Life Cycle of a Conifer

An evergreen that produces cones is called a **conifer.** As you might guess, conifers have life cycles too. The pictures show the stages in the life cycle of a conifer. Conifers are different from flowering plants. Conifers usually keep their leaves all year. Their leaves are often shaped like needles. The seeds of a conifer grow in its cones. The scales of a cone are closed tightly and keep the seeds inside dry. When the scales open, the seeds fall out of a pollinated cone.

Seeds

The brown, woody scales of a pollinated cone open, and seeds fall to the ground. Some seeds germinate and begin to grow near the parent plant. Other seeds are scattered or eaten by animals.

Seedling
A seedling grows and develops.

Growth
The tree becomes fully grown.

Cone Development and Pollination

In spring, cones begin to form on the fully grown tree. Some cones make pollen. The pollen travels by air to pollinate other cones. Seeds begin to grow in the pollinated cones.

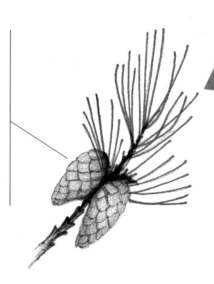

Lesson 3 Review

1. What are the three parts of a seed?

2. What are three ways seeds can be scattered?

3. Describe the life cycle of a flowering plant.

4. Describe the life cycle of a conifer.

5. **Sequencing**
   Place the following words in the correct sequence in the life cycle of a flowering plant: seedlings, seeds, germination, growth and pollination.

Chapter 1 Review

Chapter Main Ideas

Lesson 1

• Roots hold a plant tightly in soil and take water and minerals the plant needs from the soil.

• Stems hold up the leaves, flowers, and fruit of a plant and carry the water and minerals to other parts of the plant.

• Green leaves make most of the sugars a plant needs to live and grow.

Lesson 2

• When pollen reaches the center part of a flower, the flower forms seeds.

• Wind and animals can pollinate flowers.

• After a cone is pollinated, seeds form inside the cone.

Lesson 3

• A seed has several parts.

• Seeds can be scattered by wind, water, and animals.

• Seeds, germination, seedling, and growth and pollination are the stages in the life cycle of a flowering plant.

• Seeds, seedling, growth, and cone development and pollination are the stages in the life cycle of a conifer.

Reviewing Science Words and Concepts

Write the letter of the word or phrase that best completes each sentence.

a. carbon dioxide
b. conifer
c. germinate
d. life cycle
e. mineral
f. petal
g. pollen
h. pollinate
i. seed coat
j. seed leaf
k. seedling

1. An evergreen plant that produces cones is a ___.

2. The fine, yellowish powder found on the tip of some flower parts is called ___.

3. When seeds ___, the tiny plants inside begin to grow and develop.

4. All the stages in the life of a living thing are called its ___.

5. Plants need a gas in the air called ___ to make sugar.

6. When insects or birds move pollen to the center of a flower, they ___ the flower.

7. The thin, hard ___ protects the inside parts of a seed.

8. A ___ is a natural, nonliving material found in soil.

9. A young plant is called a ___.

10. A developing plant uses food stored in its ___ to grow and develop.

11. One outside part of a flower is called a ___.

Explaining Science

Draw and label a diagram or write a short answer to answer these questions.

1. How are roots, stems, and leaves important to a plant?

2. How do flowering plants form seeds?

3. What are the steps in the life cycle of a flowering plant?

Using Skills

1. Write a paragraph explaining how **sequencing** might help you understand the life cycle of a plant.

2. You see a dandelion growing in the middle of a yard of grass. What might you **infer** about how the dandelion got there?

3. Suppose you are walking through an evergreen forest on a warm, dry day. **Predict** whether you might find conifer seeds. Explain your prediction.

Critical Thinking

1. Suppose you removed the petals from a flower. **Draw a conclusion** about whether the flower could form seeds. Explain your reasoning.

2. **Predict** whether a seed can germinate without sunlight. Explain your answer.

3. You and your friend are going to plant a garden. Each of you has brought five different types of seeds to plant. **Compare and contrast** your seeds with your friend's seeds.

4. Some trees lose their leaves in cold weather. However, the trees stay alive and grow new leaves when the weather turns warm. Write a statement **inferring** how trees stay alive in cold weather.

It Just Takes Time!

You can't make an egg hatch faster. The baby chick inside needs a certain amount of time to grow. A little baby chick has to work pretty hard to come out into the world!

Chapter 2

How Animals Grow and Change

Lesson 1
How Are Babies Like Their Parents?

How do animal babies grow to look like their parents?

How do animals grow from eggs?

Inquiring about **How Animals Grow and Change**

Lesson 2
How Do Spiders and Insects Grow?

What is the life cycle of a spider?

What are the life cycles of insects?

Lesson 3
How Do Fishes, Frogs, and Mammals Grow?

What are the life cycles of fishes and frogs?

What are mammals and how do they grow?

Lesson 4
How Do Babies Learn?

What things do animals know when they're born or hatched?

How do animals learn?

Copy the chapter graphic organizer onto your own paper. This organizer shows you what the whole chapter is all about. As you read the lessons and do the activities, look for answers to the questions and write them on your organizer.

Exploring Eggs

Process Skills

- observing
- inferring
- communicating

Materials

- hard-boiled egg
- paper towel
- hand lens
- plastic knife

Explore

1. Put a hard-boiled egg on the paper towel. Tap the side of the egg on your desk to crack the shell. Carefully peel the shell away from the egg.

2. **Observe** the inside and outside of the shell with the hand lens. Record your observations.

3. Cut the egg in half with the plastic knife as shown. Observe the yolk and egg white. Make a drawing of the parts of a boiled egg. Wash your hands after handling eggs.

Reflect

1. Think of the inside of a raw chicken egg. How is the raw egg similar to a hard-boiled egg? How is it different?

2. Make an **inference.** Why is the shell important to a growing chick? **Communicate.** Discuss your ideas with the class.

? Inquire Further

How are eggs of different birds alike and different? Develop a plan to answer this or other questions you may have.

Measuring Elapsed Time

Some animals have to work for a long time to hatch from their eggs. Scientists measured the hatching times of several animals. You will learn how to find ending time and the time between events.

Remember

You can skip count by 5s to add or count on minutes.

Example 1

After a baby pigeon cracks the egg, it takes about 6 hours and 25 minutes to hatch. If a pigeon begins hatching at 5:00 A.M., what time will it finish?

Step 1

Start at 5:00 A.M.
Add the hours.

Step 2

Add the minutes.

The baby pigeon will hatch at 11:25 A.M.

Here's an example where you find the amount of time that has passed.

Example 2

Suppose a baby Japanese quail starts hatching at 9:00 P.M. and finishes at 10:35 P.M. How long does it take the baby quail to hatch?

Step 1

Start at 9:00.
Count the hours.

Step 2

Count the minutes.

It takes 1 hour and 35 minutes for the Japanese quail to hatch.

Talk About It!

How can skip counting help you count the minutes?

What's the Big Idea?

You will learn:

- how animal babies grow to look like their parents.
- how animals grow from eggs.

How Are Babies Like Their Parents?

How Cute! Young petrel birds are so fluffy and yellow. You can see how they will change as they get older. How did you look when you were a baby? Imagine how you will change as you grow.

Parents and Babies

Like plants, animals have life cycles. You recall that in a cycle, something happens over and over again. During an animal life cycle, a baby is born and then grows. When it becomes an adult, an animal can reproduce to have babies of its own kind. Then the cycle starts over again.

During its life cycle, an animal goes through several stages. Different kinds of animals go through different stages.

◀ *Like most bird babies, this young petrel looks a lot like its parents.*

When some animal babies are born, they look like their parents. For example, a lemur has the same body shape as its parents. A lemur baby also has the same number of legs as a grown lemur. The body form of a lemur remains the same throughout its life cycle.

Other animal babies look very different from their parents. Their body form changes as they go through their life cycle. When these animals become adults, they look more like their parents. Which animal babies on this page are very different from their parents?

Every living thing has a special way of living, growing, and reproducing. Scientists put animals into groups according to how they look and how they live. Some animal groups are insects, fishes, amphibians, birds, and mammals. All animals in each group grow and change in much the same way. Even so, each animal is an individual. In certain ways, each animal is different from others of its kind.

▲ *The butterfly is an insect. Like many insects, the baby looks very different from the parent.*

▲ *The frog is an amphibian. Like many amphibians, the baby will grow and change a great deal before it becomes an adult.*

Like most mammals, this lemur and its parent have the same body form. ▶

Glossary

Glossary

embryo (em′brē ō),
a developing animal
before it is born
or hatched

▲ *Insect eggs are usually very small. Insects lay eggs in water, or underground, or on wood or leaves.*

▲ *Some fish dig nests under the water. Other kinds of fish lay eggs that float on water.*

▲ *Frogs usually lay their eggs in water. The eggs stick together in a jellylike lump.*

▲ *Bird eggs come in many sizes and colors. A robin egg is fairly small. A chicken egg is a bit larger.*

Growing from Eggs

The first stage in most animal life cycles is the egg. Find the different eggs in the pictures on the left. Eggs of most animals have the same functions. Notice in the picture below that an egg protects an embryo and provides food for it. An **embryo** is a developing animal before it is born or hatched.

Different kinds of animals produce different numbers of eggs. Insects, fishes, and frogs lay many eggs. After they lay the eggs, the adults usually go away.

Inside a Chicken Egg

Thick Strings
These hold the yolk in place when the egg is turned. The hen turns the egg so it keeps evenly warm.

Embryo Spot
If the egg has an embryo, it grows on the yolk.

Air Space
The larger end of the egg has a space filled with air for the embryo.

Egg White
This liquid provides food and a cushion for the embryo.

Shell Linings
Two thin linings are just inside the shell.

Shell
The shell is hard. It has many small openings that allow oxygen and water to enter the egg. Waste gases also leave the egg through these openings.

Yolk
The yolk provides food for the embryo.

Other animals, such as birds and mammals, produce fewer eggs. The parents care for and feed their young for a while. Many parents stay until the young grow big enough to get their own food.

Each kind of animal has a special kind of egg. Some animals, such as birds, lay eggs with hard shells. Other animals, frogs for example, lay eggs in a jellylike substance. The eggs do not have hard shells. Babies of other animals, such as dogs and cats, develop inside the mother's body.

About twenty-one days after the egg is laid, the chick starts to hatch. It pokes its beak through the shell linings and into the air space. It pecks at the shell with a special egg tooth on its beak. ▶

The chick pushes hard against the shell. The shell finally breaks and the wet chick comes out. ▶

▲ *Soon the chick dries out and looks for food.*

Lesson 1 Review

1. Which kinds of animal babies change the most as they grow?

2. What are the functions of eggs?

3. **Elapsed Time**
An ostrich takes more than fifty hours to hatch. Does the hatching process take place during both daylight and nighttime?

What's the Big Idea?

You will learn:

- to describe the life cycle of a spider.
- to describe the life cycles of different kinds of insects.

The spider is near her egg sac. The baby spiders that will come out of it are sometimes called spiderlings. ▼

How Do Spiders and Insects Grow?

EEK! A bug! You may think that all crawly animals are insects or bugs, but they're not. Actually, only some insects are bugs, and spiders are not insects at all. Spiders are arachnids.

Life Cycle of Spiders

All spiders belong to a group of animals called arachnids. Insects belong to another group. Insects and arachnids are similar in some ways. They are both covered with a skeleton and have joints in their legs. You might notice differences in the body forms of the spider and insect in the pictures on these two pages. Spiders have eight legs. Insects have six legs. A spider's body has two main parts. An insect's body has three parts. Many insects have wings. Spiders do not have wings. Neither spiders nor insects have bones.

Spiders spin strong silk threads. They use the threads to make sticky webs that trap insects. A spider mother, like the one shown, makes a silk sac to store her eggs. Baby spiders grow in the egg sac. When they have grown large enough, the young spiders break the sac and come out. They have the same body form as the adults.

Life Cycles of Insects

Insects are the largest group of animals. There are about a million different kinds of insects.

Different insects have different kinds of life cycles. Most insects lay eggs. Some insects lay their eggs in a clump. Others lay single eggs. Some young insects have almost the same body form as their parents. Other young insects look very different from their parents.

Some insects, like the cockroach in the picture, have a three-stage life cycle. Other insects that have a three-stage life cycle are grasshoppers, crickets, and dragonflies.

The egg is the first stage in the cockroach life cycle shown below. The second stage is called the **nymph.** It hatches from an egg. The nymph has a covering over its body that cannot stretch as the nymph grows. The covering breaks off or is shed. Each time the covering is shed, a new one forms. The nymph sheds a covering several times as it grows. The adult is the third stage. The adult can reproduce.

A Three-Stage Insect Life Cycle

1 Egg
The egg is the first stage. An egg case like this one can hold many eggs.

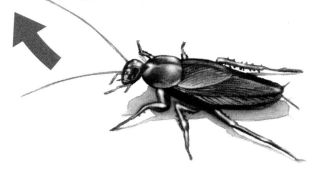

2 Nymph
A nymph hatches from an egg. A nymph has the same body form as the adult cockroach, but it does not have wings.

3 Adult
When the nymph grows to adult size, it has wings. Adults can reproduce.

The life cycle of butterflies is different from the life cycle of cockroaches. You have seen that young butterflies are very different from their parents. Butterflies have a four-stage life cycle. Ants, moths, and bees are some other insects that have a four-stage life cycle. Follow the diagram to see the different stages in the butterfly's life cycle.

A Four-Stage Insect Life Cycle

1 Egg

Butterflies usually lay their eggs on leaves. A mother butterfly finds the right kind of plant for the young to eat after they hatch.

2 Larva

Caterpillars eat a great deal. Many of them eat part of their shell as their first meal. Then the caterpillars eat leaves and other parts of plants. They grow quickly.

3 Pupa

The larva spins a silk thread that wraps around its body. The thread holds the larva in place while a hard covering forms over the insect. Inside the covering, the pupa does not eat. Its body changes form. The pupa can stay in the covering for weeks or months.

The first stage of the life cycle is the egg. The caterpillar is the butterfly's second stage. This second stage is also called a **larva**. The larva has a different shape than the adult. The larva spins a covering for itself. Inside the covering, the larva grows to become a **pupa**. After some time, the covering opens and the butterfly comes out.

Glossary

larva (lär⁄və), a young animal that has a different shape than the adult

pupa (pyü⁄pə), the stage in the insect life cycle between larva and adult

4 **Butterfly**
The covering splits open and the adult butterfly comes out. At first, the butterfly cannot fly. Its wings are damp and folded. When its wings are smooth, flat, and dry, the insect can fly away. After a time, the butterfly lays eggs and the cycle starts again.

Lesson 2 Review

1. How does the body of a young spider compare with the body of an adult spider?

2. How is a three-stage insect life cycle different from a four-stage insect life cycle?

3. **Elapsed Time**
It takes a caterpillar about two hours to hatch from its shell. Suppose it started hatching at 12:15 P.M. At what time would it finish?

A 39

Observing the Life Cycle of a Beetle

Process Skills

Process Skills

- observing

Materials

- 6 mealworms
- paper towel
- hand lens
- plastic spoon
- cornmeal or cereal flakes
- plastic jar
- piece of raw potato
- cheesecloth
- masking tape

Getting Ready

You can learn about different stages in an insect's life cycle by observing mealworms. A mealworm is the larva in the life cycle of a beetle.

You will not see the egg stage of the beetle's life cycle. They are small and difficult to see.

Follow This Procedure

1 Make a chart like the one shown. Use your chart to record your observations and drawings.

Stage	Date	Observations and drawings
Larva		
Pupa		
Adult		

2 Record the date on your chart. Place one beetle larva on a paper towel. **Observe** the larva with a hand lens (Photo A). Record and draw your observations.

3 Use the spoon to put a layer of cornmeal or cereal in the jar. Put the mealworms on the plastic spoon. Place them carefully on the food in the jar. Add a piece of raw potato to the jar for moisture.

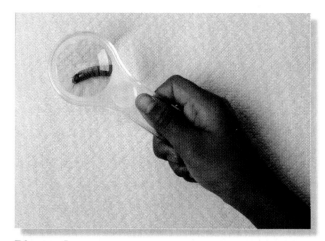

Photo A

Photo B

④ Cover the jar with cheesecloth. Use tape to hold the cheesecloth in place (Photo B).

⑤ Put the jar in a warm place away from direct sunlight.

⑥ Observe the mealworms for a short time every day. Replace the food and potato once a week.

⑦ After about two weeks you should see that some mealworms have changed to the pupa stage. Repeat step 2, observing one beetle in the pupa stage.

⑧ After a few more weeks you should see adult beetles in the jar. Repeat step 2, observing one beetle in the adult stage.

Self-Monitoring
Have I made drawings and recorded my observations for three stages of the beetle life cycle?

Interpret Your Results

1. How many days did it take for the larva to change to the pupa stage? How many days did it take for the larva to become an adult?

2. Which stage of the beetle life cycle was not observed in this activity?

3. Compare and contrast the three stages of the beetle life cycle that you observed. How are they alike and different?

 Inquire Further

What other insects have a life cycle like that of the beetle? Develop a plan to answer this or other questions you may have.

Self-Assessment

- I followed instructions to **observe** three stages in the life cycle of a beetle.
- I properly cared for the beetles.
- I recorded and drew my observations.
- I stated how long the stages in the development of the beetle lasted.
- I compared and contrasted the three stages of the beetle life cycle that I observed.

You will learn:

- about the life cycles of fishes and frogs.
- about the life cycles of mammals.

Glossary

Glossary

amphibian
(am fib′ē ən), an animal with a backbone that lives part of its life cycle in water and part on land

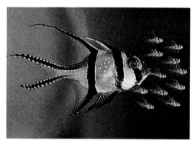

▲ *Do the baby fish look like their parent?*

Lesson 3

How Do Fishes, Frogs, and Mammals Grow?

It's true! Fishes and frogs are completely different animals. There are lots of differences between them. However, you might be surprised at how much they are really alike.

Fishes and Frogs

Fishes and frogs are two kinds of animals that have backbones. You have a backbone too. You can feel it in the middle of your back. Your backbone helps you stand straight. It also helps you bend and move.

Fishes and frogs are alike in another way. They both lay their eggs in or under water. The eggs do not have hard shells. They have thin coverings. Oxygen from the water passes easily into the egg.

Notice that the young fish in the picture look very much like the adult. In time, they will grow bigger, but their body shape will stay the same. Some kinds of fishes swim as soon as they hatch.

Frogs belong to a group of animals called amphibians. An **amphibian** lives part of its life in water and part on land. Other amphibians are toads and salamanders.

Like all animals, fishes need oxygen. Fishes get oxygen from water with their **gills.** Now compare the baby fish and the tadpole. A **tadpole** is a young frog. Does the tadpole look more like a fish or its frog parents? Like the fish, the tadpole lives in water and has gills. As the diagram shows, the young frog will grow to look like its parents. Its legs will grow and its gills will disappear. It will breathe with lungs. It will get oxygen from air.

Glossary

gills (gilz), the parts of fish and tadpoles that are used to take in oxygen from water

tadpole (tad′pōl), a very young frog or toad

Life Cycle of a Frog

1 Frog Eggs

Frogs lay many eggs at a time. The eggs contain food for the embryos.

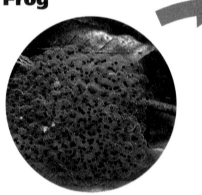

2 Young Tadpole

Frog eggs hatch in water. At first the tadpole has gills and swims about, looking for food.

3 Older Tadpole

The back legs start to grow first. Then the front legs grow. The tail becomes smaller and smaller. The lungs start to develop.

4 Adult Frog

When the frog is fully grown, it breathes with its lungs. It also gets oxygen through its skin. Some frogs live on land and some live both on land and in water.

A43

Mammals

All the animals on these two pages seem to be very different. Most of them live on land, but some live in water. Some of them swim and some of them fly. However, they all have backbones, and they all belong to the same group of animals. They are all mammals.

A **mammal** is an animal that has a backbone and hair or fur. Some mammals, such as dogs, have a great deal of fur. Some other mammals, such as whales, have very little hair. Hair helps to keep mammals warm. Mammals that live in very cold places have very thick hair. Some mammals have whiskers, which are long, stiff hairs around the nose. Whiskers help the animal feel things moving around them.

Whales are one kind of mammal that lives in water. Other water mammals are dolphins, porpoises, and manatees. ▼

A 44

All mammals need to get oxygen from air. They all have lungs. Whales and dolphins are mammals that live in water. They have to come partway out of the water to get air.

Mammal mothers produce milk to feed to their new babies. Special parts of the mother's body make milk. Only mammal mothers can produce milk. Other animals that feed their babies have to catch or gather the food.

Most mammal babies stay with one or both of their parents for a time. Some animals live in groups with many adults. The adults teach the young how to get food. The adults also help keep the young animals safe.

▲ Bats are mammals that fly. They make sounds and listen for echoes to help locate food.

▲ The duckbill or platypus is one of very few kinds of mammals that lay eggs.

◀ Kangaroo babies cannot see or hear when they are born. Their mothers carry the babies in a pouch on their bellies. Opossums grow the same way.

▲ Armadillos have bristly hair on their bellies. To protect themselves, they can curl up inside their armor bands. They eat a lot of insects.

◀ You might have a housecat as a pet. Some types of cats are wild and do not make good pets.

Mammal babies start as eggs, just like other animals. Most mammals do not lay eggs, though. The embryos grow inside the mother's body. Embryos of different kinds of mammals grow inside the mother for different amounts of time.

Different kinds of mammals have different numbers of babies at a time. Some mammals, such as whales, usually have just one baby at a time. Others, such as cats, can have four to six babies at a time. The diagram shows the life cycle of one type of cat, a bobcat.

Mammal Life Cycle

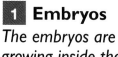 **Embryos**

The embryos are growing inside the mother's body. The mother's body is larger along the sides and around her belly.

 **Babies**

After the babies are born, the mother feeds milk to them. The mother takes care of the babies.

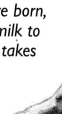

4 Adult
The adult mammal can become a parent.

3 Young Animals
Young mammals spend time with their parents. They learn many skills, such as how to get food. They eat the same food as adults.

Lesson 3 Review

1. How is the life cycle of a fish different from the life cycle of a frog?

2. What are the stages in a mammal life cycle?

3. Sequencing
Use what you know about the life cycle of frogs to put these steps in the correct order: breathe with gills; breathe with lungs; hatch from eggs.

You will learn:
- what animal babies know when they are born or hatched.
- about ways that animals learn.

Glossary

instinct (in′stingkt), an action that an animal can do without learning

Most kittens move this way when they play. Older cats use these same actions when they hunt for food. ▼

Lesson 4

How Do Babies Learn?

Swat! Did you ever see a kitten play with a toy? The kitten bats at it with its paws. No one taught the kitten below how to play that way. It's just something that kittens know how to do.

What Some Animals Know

Animals know how to do some things as soon as they are born. Most animal babies know how to get food right away. When puppies are born, they know how to get milk from their mother. Chicks know how to peck at things. Soon after caterpillars hatch, they know how to eat leaves. A baby kangaroo knows how to crawl into its mother's pouch. An action that an animal can do without learning is an **instinct.** Babies must use instincts in order to survive. They have not had time to learn these important skills.

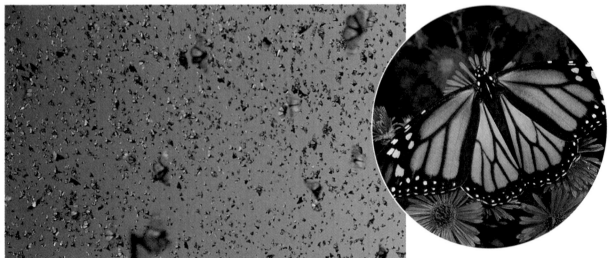

◀ Monarch butterflies fly south in late summer and early fall. In spring, they return north. They lay their eggs there in summer.

Adult animals use instincts too. Birds like the one below know how to build nests for their eggs. They protect their eggs and keep them warm. Some kinds of fishes make nests for their eggs too. Bees, ants, and other insects all use instincts when they make nests for their young.

All animals of the same kind do certain things in the same way. For example, all over the world garden spiders like the one shown build the same kind of webs.

Some animals, such as monarch butterflies, use instincts to travel great distances. Most traveling animals go toward warmer places. Others travel to look for food. Notice how the group of traveling monarchs seems to fill the sky.

▲ The shape of a web is a clue about the kind of spider that made it.

Baby birds know to hold their heads up and open their beaks to get fed. Adult birds know the right food for their babies. ▶

▲ *The bear cubs are learning to catch fish by watching their mother. Later, they will imitate her and catch fish on their own.*

Some animals live in groups, like these elephants. All the adults look after all the young animals. The young learn from all the adults. ▼

Animals that feed their babies stop feeding them after a while. Then the young animals have to learn how to get their own food. They also have to learn how to take care of themselves.

Animals learn in many ways. Some young animals learn by watching their parents. The bear cubs in the picture are learning how to catch fish. Some young birds learn to sing by imitating their parents.

Young lion cubs learn how to hunt and fight. Even before they can hunt, their mother trains them with her tail. She flicks her tail around and the cubs try to catch it. Their skills get better as they practice. Some young animals, like the elephants in the picture, might learn from other adults too.

Young animals also learn from each other. Maybe you have seen chimps like these playing in a zoo, but they seem to just be having fun. They are also learning some important things. Friendly fights among young animals help them learn about one another. They learn which animals are gentle and which are not. The young animals are learning how to get along with others.

Animals can learn by doing things by themselves. A bird might fall down when it first starts to fly. It learns to do better by practicing. Sometimes young animals learn by making mistakes. For example, a young bird might eat a certain kind of caterpillar. If that caterpillar tastes awful, the bird learns to not eat that kind again. Think about things that you learned to do by practicing.

▲ These chimps are learning to get along with others as they grow up.

Lesson 4 Review

1. What is an important thing an animal must be able to do as soon as it's born or hatched?

2. What are some ways that animals learn?

3. **Sequencing**
   Bears are mammals. Use what you know about mammals to put these steps in the correct order: drinks milk from its mother; teaches young bears to hunt; catches fish on its own to eat.

Chapter 2 Review

Chapter Main Ideas

Lesson 1
• Adult animals have the same body form as their parents even though some animals have a different body form when they are young.
• Most animals go through stages of growth, starting as eggs.

Lesson 2
• During a spider's life cycle, its body form does not change.
• Some insects have a three-stage life cycle and others have a four-stage life cycle.

Lesson 3
• During a fish's life cycle, its body form does not change. During its life cycle, a young frog first looks like a fish and lives in water.
• Mammals are animals with backbones that have hair or fur, and the mothers produce milk.

Lesson 4
• Some animal babies know how to get food when they are born.
• Animals learn how to do things as they grow older.

Reviewing Science Words and Concepts

Write the letter of the word or phrase that best completes each sentence.

a. amphibian   f. mammal

b. embryo   g. nymph

c. gills   h. pupa

d. instinct   i. tadpole

e. larva

1. An insect at the stage in its life cycle when it looks like an adult but has no wings is a ____.

2. A developing animal before it is born or hatched is an ____.

3. The stage in the insect life cycle between the larva and the adult is the ____.

4. A young insect that has a different shape than the adult is a ____.

5. An animal with a backbone that lives part of its life cycle in water and part on land is an ____.

6. Fishes and tadpoles have ____ to take in oxygen from water.

7. A ____ is an animal with a backbone and hair.

8. A very young frog or toad is a ____.

9. An ____ is an action that is not learned.

Explaining Science

Draw and label a diagram or write a paragraph to answer these questions.

1. What happens during an animal's life cycle?

2. How is the life cycle of a cockroach different from the life cycle of a butterfly?

3. How are fishes, frogs, and mammals alike and how are they different?

4. What are some things that some adult animals do for their young?

Using Skills

1. Suppose a goose starts hatching at 11:15 A.M. It is completely hatched at 3:15 A.M. on the following day. Find the **elapsed time** for the hatching process.

2. Suppose that you are taking a nature walk with your class. In a pond, you see many small animals swimming about. The animals look like small fish, but they have tiny back legs. What might you **infer** about the animals swimming in the pond?

Critical Thinking

1. Matt's pet cat is about to have kittens. Matt cannot keep all the kittens in his home. He is trying to find people who will give good homes to the kittens after they are born. **Explain** to Matt why the kittens need to stay with their mother for a time before they can be moved to a new home.

2. You might see many caterpillars at certain times of the year. **Draw a conclusion** about why some people might be worried about trees at those times of the year.

3. All birds belong to one group of animals. **Observe** several kinds of birds. Make a list of the ways that the birds are alike.

Home Sweet Home!

Imagine a bird living at the bottom of an ocean. You know this example wouldn't work. The location isn't strange, but every living thing must live in its own special place to survive.

Chapter 3

Living Things and Their Environments

Inquiring about **Living Things and Their Environments**

Lesson 1
Where Do Organisms Live?

What is a habitat?

How can organisms change their environments?

Lesson 2
How Are Organisms Adapted to Their Environments?

What is an adaptation?

How can adaptations help organisms meet their needs?

Lesson 3
How Do Organisms Get Food?

What are producers and consumers?

What is a food chain?

Lesson 4
How Do Organisms Live Together?

How can the size of a population change?

What makes up a community?

How can increasing populations affect an environment?

Copy the chapter graphic organizer onto your own paper. This organizer shows you what the whole chapter is all about. As you read the lessons and do the activities, look for answers to the questions and write them on your organizer.

Exploring Where Pond Snails Live

Process Skills

- observing
- inferring

Materials

- 2 elodea plants
- plastic bottle with water
- plastic spoon
- pond snail
- plastic cup
- hand lens
- colored pencils or crayons

Explore

1 Make a bottle aquarium for a pond snail. Place two elodea plants in the bottle of water.

2 Use the plastic spoon to carefully move the snail from the plastic cup to the bottle of water. Replace the cap on the bottle.

3 **Observe** the snail with the hand lens. Record a description of the snail. Make a drawing of the snail and its home.

4 Place the bottle in a lighted place but not in direct sunlight. Observe the bottle for a few weeks. You should not have to add food, water, or air to the bottle. Try to observe the snail as it is eating. Record your observations.

Reflect

Make an **inference.** Did the snail have all it needed to survive in the bottle? Explain.

? Inquire Further

What conditions would a land snail need to survive? Develop a plan to answer this or other questions you may have.

Comparing Numbers

Slurrrp! Some tortoises love to eat snails and slugs. They can eat a lot of them in a lifetime because they live to be so old.

Which can live longer, a box tortoise or a Marion's tortoise? You can **compare** numbers to find out.

Example 1
Compare 152 and 138.

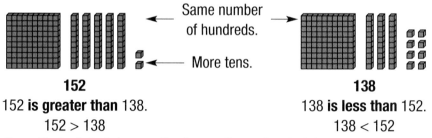

Same number of hundreds.

More tens.

152

152 **is greater than** 138.

152 > 138

138

138 **is less than** 152.

138 < 152

So, a Marion's tortoise can live longer than a box tortoise.

You can also compare numbers using place value.

Example 2
Compare 1,550 and 1,830.

Step 1	Step 2
Begin at the left. Compare. **1,**550 *Both numbers have* **1,**830 *1 thousand.*	Find the first place where the digits are different. Compare. 1,**5**50 *5 hundreds is less* 1,**8**30 *than 8 hundreds.*

So, 1,550<1,830 or 1,830>1,550.

Talk About It!

Since 5 is greater than 2, is 5,360 greater than 26,314? Explain.

Math Vocabulary

compare (kəm per´), to decide which of two numbers is greater

Number of Years Tortoises Can Live	
Marion's tortoise	152 years
Box tortoise	138 years

Math Tip

You can use these steps to compare numbers with any number of digits.

You will learn:
- what a habitat is.
- how organisms can change their environments.

Many people live in a city habitat, such as Houston, Texas. ▼

Lesson 1

Where Do Organisms Live?

A squirrel chatters noisily as it climbs the nearest tree. A bird chirps loudly. Ants and other insects busily crawl along the ground. Why do so many things live around here?

Habitats

Think of all the living things, or **organisms,** around you. Organisms need food, water, air, shelter, and space to live. An organism gets everything it needs from its habitat. A **habitat** is the place where an organism lives. Notice the city habitat shown here. What is the habitat like where you live? You probably get food from the store. Think about where you get water and shelter. Everything you need to live is in your habitat.

Forests, such as the one in the picture, are one of the many different kinds of habitats in the world. Forests are habitats that get a lot of rain and have many trees. Plants that live in a forest habitat need plenty of water. Trees, bushes, moss, and grass live in a forest habitat. Animals that live in a forest need the food, water, and shelter a forest habitat provides. Birds, such as the one shown below, live in a forest habitat. They make their nests in trees, shrubs, and tall grass. They drink water from nearby ponds and eat insects and seeds. Think of other plants and animals that live in a forest habitat.

There are many different kinds of forests. ▼

This grosbeak gets everything it needs to live from its forest habitat. ▶

Certain organisms live in each kind of habitat. Remember that in order for organisms to survive, they must get everything they need from their habitat. On these two pages, read about different habitats and the organisms that live there.

A Rain Forest Habitat

This parrot's habitat is in the trees of the rain forest, where there is plenty of sun and rain. Parrots fly from tree to tree to find the buds, fruit, berries, leaves, and nuts they eat. Parrots find almost everything they need in the trees of the rain forest and very rarely go to the ground. ▶

An Ocean Habitat

Most of the surface of the earth is covered by oceans. The animal below, called an eel, is one organism that lives in the ocean. Many eels live in rocky areas of the ocean and hide in the small, narrow holes of the rocks. They mostly eat other fish. ▶

◀ A Wetland Habitat

Painted turtles live in the shallow, fresh waters of ponds, lakes, and streams. They eat the snails, insects, and crayfish found there. They also eat the plants that grow in or near the water. Painted turtles often lie on top of a log or by the edge of the water to warm themselves in the sun.

A Desert Habitat

Kangaroo rats live in the desert. They dig a hole in the ground for shelter. Many deserts are very hot during the day and become very cool at night. Kangaroo rats sleep during the day and search for food at night. Deserts are also very dry. Usually, kangaroo rats do not drink water. They get all of the water they need from the seeds they eat. ▼

Glossary

environment
(en vī′rən mənt), all the things that surround an organism

How Organisms Can Change Their Environments

All the things that surround an organism are called its **environment.** Organisms can change their environments in different ways. The beaver in the picture uses its strong front teeth to cut down trees. It uses the trees to make a dam. The dam blocks the water of a stream, creek, or river. A small pond forms. As the beaver makes the dam in the picture higher, the pond may widen into a lake.

The beaver has changed its environment. The stream may dry up because it is blocked by the dam. There is a lake where there was no lake before.

The beaver lives in a building called a lodge. It cuts down trees to build its lodge at the edge of the water. ▼

After the beaver has built its lodge, it cuts down more trees to build a dam. The dam may widen the water into a lake. The lake helps protect the beaver from its enemies. ▼

Another animal that changes its environment is the groundhog in the picture. Notice the tunnels that groundhogs dig underground. They live in these tunnels, or burrows. They may dig more than one hole in the ground to enter the burrow.

Groundhogs like to live near open areas. Groundhogs often live in the same areas as people, which can be a problem for both. Groundhogs will dig tunnels in backyards and gardens, especially if there are open areas nearby. Sometimes people try to get rid of groundhogs by filling in their holes. However, many times, the groundhogs just dig other holes in the same area.

A groundhog and a groundhog burrow ▼

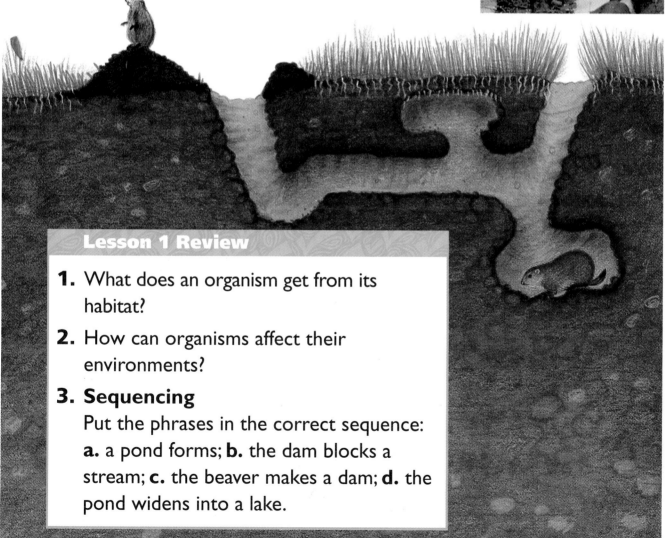

Lesson 1 Review

1. What does an organism get from its habitat?

2. How can organisms affect their environments?

3. **Sequencing**
   Put the phrases in the correct sequence:
   a. a pond forms; **b.** the dam blocks a stream; **c.** the beaver makes a dam; **d.** the pond widens into a lake.

Comparing How Quickly Water Moves Through Leaves

Process Skills

- observing
- inferring

Process Skills

Materials

- safety goggles
- pencil
- 2 index cards
- cutting from a jade plant
- cutting from a coleus plant
- 4 clear plastic cups
- water
- masking tape

Getting Ready

In this activity you will find out how quickly water moves through the leaves of different plants.

Follow This Procedure

1 Make a chart like the one shown. Use your chart to record your observations.

Day	Jade plant observations	Coleus plant observations
1		
2		
3		
4		
5		

2 Put on your safety goggles. Use the pencil to make a hole in the center of each index card.

3 Place the stem of the jade plant through the hole of one index card. Place the stem of the coleus plant through the other index card.

4 Fill two plastic cups half full with water.

⚠ **Safety Note** *Wipe up any spilled water immediately.*

5 Place one index card on top of each cup (Photo A). Make sure the two plant stems are in the water. Tape the index card to the cup.

6 Place an upside-down empty cup over each plant. Tape it to the index card (Photo B). Wash your hands after handling the plants.

Photo A

Photo B

Self-Monitoring
Did I set up the experiment correctly?

7 Put the two plants in the same place. Do not put the plants in direct sunlight. **Observe** the top cups every day for five days. Record your observations.

Interpret Your Results

1. Which top cup contains more water after five days?

2. Draw a conclusion. Does water pass faster through jade leaves or coleus leaves? Explain.

3. Make an **inference.** Which plant do you think needs to live in an environment with lots of moisture? Explain.

**Inquire Further**

How quickly does water move through the leaves of other kinds of plants? Develop a plan to answer this or other questions you may have.

Self-Assessment

- I followed instructions to find out how quickly water moves through leaves.
- I **observed** the plants and cups for five days.
- I recorded my observations.
- I drew a conclusion about how fast water moves through leaves of a jade and a coleus plant.
- I made an **inference** about the environment of a plant.

Glossary

adaptation
(ad′ap tā′shən), a structure or behavior that helps an organism survive in its environment

Lesson 2

How Are Organisms Adapted to Their Environments?

A fish stares back at you from its watery home. Its mouth keeps opening and its fins keep waving. How can a fish live in all that water?

Adaptations

All organisms have adaptations. An **adaptation** is a structure or behavior that helps an organism survive in its environment. Some organisms have adaptations that help them live on land. What would be one adaptation that would help an organism live on land? Other organisms have adaptations that help them live in water. Think of one adaptation for living underwater.

Some organisms have adaptations for living in a cold area. For example, mountain goats such as this one live where it can get very cold. They have very thick fur with a layer of long hair on top. These two layers keep the goats warm during freezing weather.

◄ *A mountain goat is adapted to live in the mountains.*

A 66

Mountain goats also have adaptations that help them live in steep, rocky mountain areas. The toes on their hooves can spread out wide to grip large areas of rock. The toes also can close together firmly to grasp narrow edges of rock. Mountain goats have pads on the bottom of their hooves that prevent skidding or slipping. What might be one behavior a mountain goat has that helps it survive?

Gills and fins are two adaptations that help fish live underwater. Instead of having lungs as you do, fish have gills. A fish takes in water through its mouth. The water passes over the fish's gills. Oxygen goes from the water into the blood in the fish's gills.

Notice the fins and tail on the fish. A fish moves its tail to go forward in the water. It moves its fins to stay straight up and down. Fish also use their fins to start moving, slow down, change direction, and stop.

A fish is adapted to live in the water. ▼

▲ Texas Horned Lizard

Can you see the animal in this picture? The colors of some animals match the color of their surroundings. This makes it difficult for the animal to be seen. The Texas horned lizard blends into its surroundings for protection against enemies. Think how the lizard's color might help it catch animals for food.

Adaptations Meet Needs

Adaptations help organisms meet their needs. Remember that animals need food, water, air, and shelter. Plants need air, water, sunlight, and space to grow. Many living things need protection from animals that might eat them. Organisms have many different kinds of adaptations. Read about the organisms on these two pages. You may have seen some of these organisms. However, you might not be familiar with their adaptations.

Giraffes

Adaptations help a giraffe get food and protect it from enemies. A giraffe often eats the top parts of a tree. Think of how the structure of the giraffe might help it get food. Its color pattern makes the giraffe hard to see when it stands among trees. This color pattern may help protect the giraffe from enemies. ▼

▲ Hover Fly

Harmless animals sometimes look like animals that are poisonous or sting. This insect, called a hover fly, looks like a bee. A bee stings, but a hover fly does not sting. However, because a hover fly looks like a bee, enemies think it is dangerous and stay away.

▲ Redwood Trees

Redwood trees are evergreens. They grow on the west coast of the United States. They are thought to be the tallest trees in the world. The tallest redwood is taller than the Statue of Liberty. The very thick bark of a redwood tree is an adaptation. Redwood trees do not burn because the thick bark protects them from fire. The bark also helps protect redwoods from insects, disease, and decay.

Lesson 2 Review

1. What is an adaptation?

2. Explain how the structure of a giraffe would help it get food.

3. Compare Numbers
Wings on birds are an adaptation. A hummingbird flaps its wings about 4,200 times a minute. A chickadee flaps its wings about 1,620 times a minute. Which bird flaps its wings more times a minute?

What's the Big Idea?

You will learn:
- the difference between producers and consumers.
- the parts of a food chain.

Glossary

Glossary

producer
(prə dü′sər), an organism that makes its own food

consumer
(kən sü′mər), an organism that eats food

◀ *This prairie dog is a consumer. It eats grasses and other plants.*

Lesson 3

How Do Organisms Get Food?

Oh, **yuck!** That frog just ate a fly! Why do frogs eat such creepy stuff? You like peanut butter and banana sandwiches. Should all organisms eat what you do? Or maybe there is more to food than just taste?

Producers and Consumers

How do organisms get the food they need to live? Plants and animals get food in different ways. Remember that green plants use energy from the sun to make sugar. A green plant is a producer. A **producer** is an organism that can make its own food.

Most organisms, such as the prairie dog in the picture, cannot make their own food. They are called **consumers.** Consumers must eat food. Which organisms in the picture on the next page are producers? Which organisms are consumers?

Different consumers eat different kinds of food. Remember that the frog and you eat different foods. Some consumers, such as rabbits, deer, horses, and elephants, eat only plants.

Other consumers, such as hawks, spiders, cats, and snakes, eat only animals. Still other consumers, such as bears, foxes, and raccoons, eat both plants and animals. Do you eat plants, animals, or both?

Both producers and consumers live on the prairie. ▼

Deer

Tall Grasses

Owl

Goldenrod

Coyote

Quail

Clover

Glossary

food chain, the way food passes from one organism to another

A Food Chain

Organisms depend on each other for food. Use your finger to follow the arrows connecting the pictures on these two pages. The mouse is eating grass. A hungry snake slithers by and swallows the mouse. An owl out hunting for food catches and eats the snake. The grass, mouse, snake, and owl are all part of a food chain. A **food chain** is the way food passes from one organism to another. The grass makes sugars. The mouse eats the grass. The snake eats the mouse. The owl eats the snake.

A food chain ▼

Grass
The green grass produces sugars.

Mouse
The mouse eats the grass.

All food chains begin with producers. A plant is a producer. Remember that plants use energy from sunlight to produce sugar. All consumers depend on producers for food. Animals are consumers. Some consumers, such as the mouse, eat plants. Other consumers, such as the snake and owl, are predators. A **predator** is an animal that captures and eats other animals. Animals that are captured and eaten by other animals are called **prey.**

Every living thing—producer, predator, or prey—is a link in a food chain. Food chains are found in the soil, on land, and in the water.

Glossary

predator (pred′ə tər), an organism that captures and eats other organisms

prey (prā), an organism that is captured and eaten by another organism

Snake
The snake eats the mouse.

Owl
The owl eats the snake.

Lesson 3 Review

1. What is a producer? What is a consumer?

2. Give an example of a food chain.

3. **Compare Numbers**
   A lion can run 80 kilometers an hour to catch a gazelle. The gazelle can run 76 kilometers an hour. Compare the numbers. Which animal runs faster?

You will learn:

- how the size of a population can change.
- what makes up a community.
- how increasing populations affect an environment.

Lesson 4

How Do Organisms Live Together?

Look up! That group of geese is called a flock. A group of kangaroos is a mob. Whales form a pod. Some animals have different group names, but one word describes them all.

Populations

The gazelles shown here form a population. A **population** is a group of organisms of the same kind that live in the same place at the same time.

The size of a population can change. Populations often increase when there is enough food, water, and space to live. Populations often decrease when there is not enough food, water, or space. Think of how the amount of sunlight and water might cause the size of this water lily population to change. A population will grow smaller when more members die than are born. A population will grow larger when more members are born than die.

Gazelles ▼

◀ *Water lilies*

Communities

Notice the different plant and animal populations around the water in the picture. Most places have more than one kind of population. Populations of cattails, turtles, and fish live in a pond. Populations of birds, trees, butterflies, and rabbits live in a forest.

All the populations that live and interact in the same place make up a **community.** There are predators, prey, and food chains in a community. Living things in a community depend on each other for food and shelter.

Think about the city or town where you live. What other populations live there? Think about how the different populations in your community interact.

Glossary

community
(kə myü′nə tē), all the plants, animals, and other organisms that live and interact in the same place

A community ▼

Increasing Populations

Hungry deer can destroy their own environment by eating seeds, seedlings, and parts of plants. In winter, they can also destroy the land searching for grass under the snow. ▼

How can increasing populations affect the environment? In some areas, deer such as the one shown have very few predators. As a result, the deer population increases. Deer eat the twigs, leaves, and bark of trees. They eat tree seedlings which would have grown into new trees. They also eat bushes, grass, and flowers. A large herd of hungry deer can destroy any area where these plants grow.

Kudzu, the weedy vine in the picture, grows quickly. It climbs trees, power poles, automobiles, and anything else in its way. Kudzu can destroy forests by growing up the trees and blocking out the sunlight.

▲ *Kudzu has been used for food and in medicine for many years in China and Japan. In the United States, kudzu is used for shade and to prevent soil from washing away. Kudzu grows so fast and causes so much harm, it is now considered a weed.*

Lesson 4 Review

1. What can cause the size of a population to change?

2. What makes up a community?

3. How can increasing populations cause damage?

4. **Compare Numbers**
Populations of cockroaches and dragonflies live in communities. Scientists know of 5,000 kinds of dragonflies and 3,700 kinds of cockroaches. Which has more kinds?

Experimenting with a Plant Habitat

Materials

- safety goggles
- marker
- 3 paper cups
- sharp pencil
- newspaper
- metric ruler
- plastic spoon
- seed starter mix
- radish seeds
- 3 graduated plastic cups
- tap water
- low and high concentrations of salt water solutions
- grid paper

Process Skills

- formulating questions and hypotheses
- identifying and controlling variables
- experimenting
- collecting and interpreting data
- communicating

Process Skills

State the Problem

Plants are adapted to certain conditions in their habitats. Does the amount of salt in water in the habitat affect the germination and growth of radish plants?

Formulate Your Hypothesis

If you increase the amount of salt in water given to radish seeds, will their growth be better, be worse, or not be affected? Write your **hypothesis.**

Identify and Control the Variables

The concentration of salt in water is the **variable** you can change. Add tap water to the seeds in cup 1. This will serve as a control. Add a low concentration of salt water to the seeds in cup 2. Add a high concentration of salt water to the seeds in cup 3. Plant the same number of seeds in each cup. Keep the temperature and light the same for the 3 cups of seeds.

Continued ➡

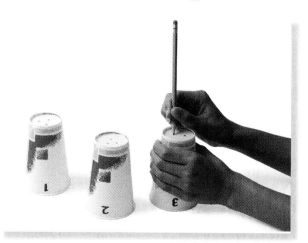

Photo A

Photo B

Test Your Hypothesis

Follow these steps to perform an **experiment**.

1 Make a chart like the one shown on page A79. Use your chart to record your data.

2 Put on your safety goggles. Using the marker, number the cups *1*, *2*, and *3*.

3 Use the pencil to make three holes in the bottom of each cup (Photo A).

4 Cover your desktop with newspaper. Use the plastic spoon to place seed starter mix in the cups, leaving 2 cm of space at the top of each cup (Photo B).

5 Plant 10 radish seeds in each cup. Cover them with a thin layer of seed starter mix.

6 Use a graduated cup to pour 60 mL of tap water into cup 1. Use another graduated cup to pour 60 mL of low salt solution into cup 2. Use the third graduated cup to pour 60 mL of high salt solution into cup 3. Repeat this step every day.

 Safety Note *Wipe up any spilled water immediately. Wash your hands after you have planted the seeds. Keep the seeds away from your mouth.*

7 Place cups 1, 2, and 3 in a sunny place. After three days, count the number of plants that have started growing in each cup.

8 **Collect data** by recording the numbers in your chart. Record any other observations you make.

9 Repeat your observations after six days and again after ten days.

Collect Your Data

	Number of plants growing			
	Day 3	Day 6	Day 10	Other observations
Cup 1 Tap water				
Cup 2 Low salt solution				
Cup 3 High salt solution				

Interpret Your Data

1. Label your grid paper as shown. Use the data from your chart to make a bar graph. Graph only the number of plants growing after ten days.

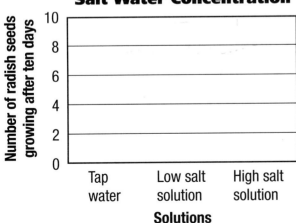

2. Study your graph. State how many seeds in each cup were still growing after ten days.

State Your Conclusion

Communicate your results. Draw a picture and describe the plants in each cup.

Inquire Further

How would other plants grow in the same salt concentrations that you used in this activity? Develop a plan to answer this or other questions you may have.

Chapter 3 Review

Chapter Main Ideas

Lesson 1
• Organisms get the things they need to live from their habitats.
• Organisms can change their environments to meet their needs.

Lesson 2
• Adaptations are structures or behaviors that help organisms survive in their environments.
• Adaptations help organisms meet their needs.

Lesson 3
• An organism can be a producer or a consumer.
• A food chain is the way food passes from one organism to another.

Lesson 4
• Kinds of organisms make up populations, and the size of a population can change.
• Different populations of organisms that live in the same place form a community.
• Increasing populations can affect an environment.

Reviewing Science Words and Concepts

Write the letter of the word or phrase that best completes each sentence.

a. adaptation   **g.** organism

b. community   **h.** population

c. consumer   **i.** predator

d. environment   **j.** prey

e. food chain   **k.** producer

f. habitat

1. An animal that is captured and eaten by another animal is called ___.

2. An ___ is a living thing.

3. A ___ captures and eats other living things.

4. A ___ is made up of all the populations that live and interact in the same place.

5. A living thing that must eat food is called a ___.

6. All the things that surround a living thing are called its ___.

7. An ___ can help an organism survive in its environment.

8. Food passes from one organism to another in a ___.

9. The place where an organism lives is called its ___.

10. A group of organisms of the same kind, or a ___, live in the same place at the same time.

11. An organism that makes its own food is called a ___.

Explaining Science

Draw and label a diagram or write a short answer to answer these questions.

1. What do living things get from their habitats?

2. How do adaptations help organisms in their environment? Give an example.

3. What is the difference between a producer and a consumer?

4. How can a population in a community change?

Using Skills

1. An opossum sleeps about 133 hours a week. A koala sleeps about 154 hours. **Compare** the numbers. Which animal gets less sleep each week?

2. **Observe** the different plant and animal populations in your community. Make a list of your observations.

3. Suppose that all of the plants in an area die. **Infer** how this would affect the food chain in that area.

Communicate your thoughts by writing one or two sentences.

Critical Thinking

1. Put the following steps in a food chain in the correct **sequence:** a grasshopper eats the plant, a raccoon eats the frog, a plant makes sugar, a frog eats the grasshopper.

2. Wolves are predators that sometimes hunt deer as prey. Suppose the wolf population decreased or disappeared. **Predict** what would happen to the deer population.

3. Suppose it has not rained and has been very dry for a long time. What would you **infer** might happen to the plant population?

4. **Draw a conclusion** about why a food chain begins with producers.

Can They Return?

Did you ever see pictures of forests that were burned by fires? Did you think that nothing could ever live there? You might be surprised to learn that things actually were living under the ashes. And look at what happened after a few years! A fresh, new forest bloomed!

After Forest Fire

Regrowth

Chapter 4
Changing Environments

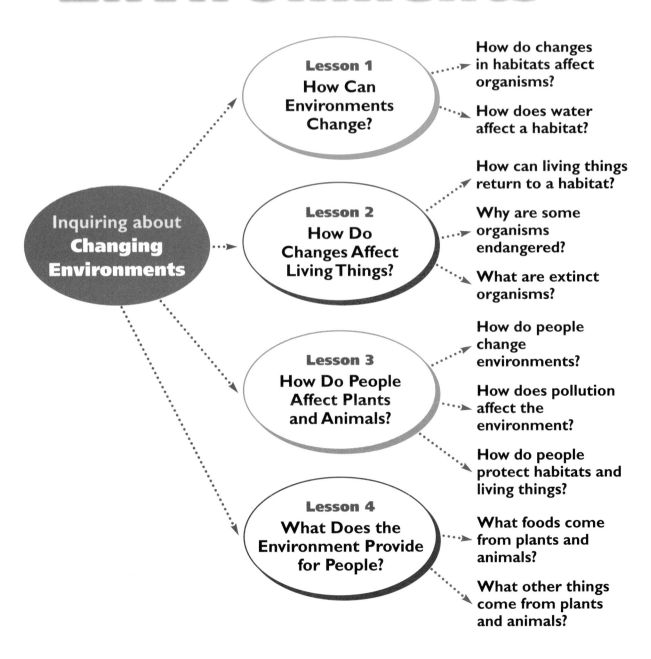

Inquiring about Changing Environments

Lesson 1
How Can Environments Change?

- How do changes in habitats affect organisms?
- How does water affect a habitat?

Lesson 2
How Do Changes Affect Living Things?

- How can living things return to a habitat?
- Why are some organisms endangered?
- What are extinct organisms?

Lesson 3
How Do People Affect Plants and Animals?

- How do people change environments?
- How does pollution affect the environment?
- How do people protect habitats and living things?

Lesson 4
What Does the Environment Provide for People?

- What foods come from plants and animals?
- What other things come from plants and animals?

Copy the chapter graphic organizer onto your own paper. This organizer shows you what the whole chapter is all about. As you read the lessons and do the activities, look for answers to the questions and write them on your organizer.

Exploring Habitats

Process Skills

- observing
- predicting
- inferring

Materials

- small jade plant cutting in pot
- elodea plant in water
- 2 plastic cups
- water
- potting soil

Explore

❶ Compare and contrast the jade plant and the elodea plant. Record your **observations.** Describe the habitats of the two plants.

❷ What would happen if the jade plant was placed in the elodea's habitat? What if the elodea was placed in the jade plant's habitat? Record your **predictions.** Explain why you predicted as you did.

❸ Fill one plastic cup with water so that it is more than half full. Put the jade plant under the water.

❹ Put potting soil in the other cup. Plant the elodea in the potting soil cup. Observe the plants every day for seven days. Record and draw your observations.

Reflect

1. How do your predictions compare with your results?

2. Make an **inference.** Which plant could live in a pond? Which plant could live in a dry area? Explain.

? Inquire Further

What would happen if you changed the amount of light in a plant's habitat? Develop a plan to answer this or other questions you may have.

Making a Prediction

In the activity, *Exploring Habitats,* you are asked to predict what would happen if the jade plant is placed in the elodea's habitat and if the elodea is placed in the jade plant's habitat. Then you will see if your prediction was correct. You make a **prediction** when you form an idea about what will happen, based on some evidence that you have.

Reading Vocabulary

prediction
(pri dik′shən), form an idea about what will happen, based on evidence

Example
In the chart below are some of the situations that you will be asked to make predictions about in Lesson 1, *How Can Environments Change?* The chart also includes a place for you to write in your predictions and also the evidence that you used to make these predictions. Make this chart on your own paper and fill it in as you read this lesson.

Situations	Prediction	Evidence
What changes might occur in a habitat?		
What will happen to plants in a flooded area?		

Talk About It!

1. What is the difference between a guess that something will happen and a prediction?

2. Predict what will happen to plants that do not get enough water.

You will learn:

- how changes in habitats affect organisms.
- how water affects habitats.

How Can Environments Change?

Great! You're going on a camping trip! Packing for the trip is a lot of work. Besides food and water, what are some other things you'll need at your campsite?

Changes in Habitats

Suppose you are going on a camping trip in a forest like the one where the deer in the picture lives. Even if you will be camping just overnight, you know you need to take certain things. Food, water, a flashlight, and blankets are just a few of the things you might need to take or be able to get at your campsite. You also will need some kind of shelter. As long as you have all the necessary things, you should be able to live comfortably at your campsite. Predict some changes that might happen at your campsite habitat. Would those changes make you want to leave?

◀ *A forest is a habitat for many living things.*

Remember that the animals that live in the forest get everything they need from their habitat. Sometimes, though, habitats change. Some changes happen quickly. A forest fire can cause a quick change. Other changes, such as not enough rain, take place over a longer time. Either way, some organisms may not be able to live in the changed habitat. Some animals leave. They find new homes somewhere else. Other animals cannot find new homes. Without a habitat, they might die.

Look at the fire shown in the picture. All the trees and materials on the ground are being destroyed. After the fire, most of the trees are burned. Some animals that live in this habitat, such as deer, will not be able to find food or shelter here. They will no longer have this habitat. To survive, they will have to leave.

▲ Some forest fires are started by lightning. Fires can spread quickly through the trees.

A forest fire can destroy communities of plants and animals. ▼

Glossary

drought (drout), a long period of dry weather

During a drought, soil becomes very dry. Do you think plants could grow here?

Many living things lose their habitat when land is flooded. ▼

How Water Affects Habitats

A living thing gets the water it needs from its habitat. Sometimes, though, an area does not have enough water for the plants and animals that live there. A long period of dry weather without enough rain is a **drought.** In a drought, lakes and rivers might dry up. The picture shows what might happen to the soil during a long drought. Notice that the upper part of the soil cracks and crumbles.

If plants do not get enough water, they might die. When plants die, the animals that usually eat them have no food. Some animals move away to search for food. Other animals might become ill or die.

Earth Science

A long drought affected a large area of the United States in the 1930s. It lasted for seven years. The area became known as the Dust Bowl. It included much of Texas, Oklahoma, Kansas, Colorado, and New Mexico.

MFA OIL

MFA

The Dust Bowl area had many farms. The photo on the right shows what happened to one farm. Crop plants could no longer grow in the dry soil. The farmer could no longer live there and none of the other farmers in the area could live there either. All the farm families had to move away.

Sometimes an area gets too much rain. The ground cannot hold all the water. Rivers and lakes overflow, and flooding occurs. The people who lived in the area shown in the picture on the opposite page lost their homes in a flood. Animals that lived there also lost their habitat. You can see the tops of trees. What do you think happened to the other plants that lived there?

Drought caused a family of farmers to leave this habitat. ▼

Lesson 1 Review

1. What can happen to organisms when a habitat changes?

2. How do changes in amounts of water affect habitats?

3. Predict
   What will happen in a flooded area after the water goes away?

What's the Big Idea?

You will learn:

- how some organisms return to a changed habitat.
- why some organisms are endangered.
- about some extinct organisms.

How Do Changes Affect Living Things?

So many birds! Have you ever seen large flocks of birds? Can you imagine a time when those birds might not exist at all? They would be gone forever. Could that ever happen?

How Organisms Can Return to a Changed Habitat

Think about the picture of the burned forest on page A87. Could you see any living things there? Would you be surprised to see any plants growing again in the burned soil?

Actually, the forest floor has many seeds that can grow. Some of them have outer coatings that do not burn easily. Others were covered by soil or were below the flames. These seeds can sprout after a fire. Notice the new plants growing in the picture on the opposite page.

Seedlings like the one in the picture sprout all the time in a forest. Many of them cannot grow very much because tall trees shade them. During a fire, the tall trees might burn. Then the seedlings will have a chance to grow into new trees. In time, new forests form. Animals return to live there.

This young seedling needs sunlight in order to grow. In a forest, many seedlings do not get the sunlight they need. ▼

This forest is growing back after a fire. Some time ago, you could not see any plants here.

Glossary

endangered
(en dān′jərd)
organism, a kind of living thing of which very few exist and that someday might not be found on the earth

Endangered Plants and Animals

You have seen that some living things might not survive when their habitat changes. When many living things of the same kind die, that kind of organism may become endangered. **Endangered organisms** are kinds of living things that are very few in number. Someday, organisms like those shown below may not be found on the earth.

Silversword

Silverswords grow only on volcanoes in small areas of Hawaii. Each plant has flowers only once in its lifetime of about fifty years. The insects that pollinate the flowers are sometimes eaten by other insects, so the plants produce very few seeds. Animals often eat the few plants that grow. ▼

▲ Hawaiian Monk Seal

Hawaiian monk seals live on and around the small islands of Hawaii. Years ago, hunters killed many seals. They are not being hunted now, but many of them do not get enough to eat. Seals eat fish and lobsters, which other animals and people also eat.

◀ Giant Panda

Giant pandas have become a symbol of endangered animals. Most of their habitat in the forests of China has been destroyed. The pandas eat parts of bamboo trees. Many of these trees have been cut down. Today, very few wild giant pandas survive.

Extinct Plants and Animals

Some kinds of plants and animals no longer live on the earth. They are **extinct organisms.** Many different events caused them to disappear.

You know that dinosaurs are extinct. Some scientists think that dinosaurs are extinct because the weather changed their habitat millions of years ago. Other scientists disagree. They are still trying to find out what happened to the dinosaurs. Scientists know why some of the organisms shown here disappeared. They are not sure about others.

Woolly Mammoth ▶
This type of elephant lived when the earth was very cold. They might have become extinct when the earth became warmer. People also might have hunted them.

▲ **Passenger Pigeon**
People hunted these birds for food and feathers. Other passenger pigeons died when their forest habitat disappeared. The last passenger pigeon died in 1914.

▲ **Giant Horsetail**
*This picture shows a **fossil** of this extinct plant. The plant was pressed into mud, which hardened into stone. Giant horsetails probably became extinct as the weather changed.*

Lesson 2 Review

1. How can organisms live once again in a habitat that has been harmed?

2. Why are some organisms endangered?

3. Why have some organisms become extinct?

4. **Predict**
   Suppose no one would catch fish or lobsters near the island habitats of Hawaiian monk seals. Predict what might happen to the number of monk seals.

You will learn:

- how people change environments.
- how pollution affects the environment.
- ways that people protect habitats and living things.

How Do People Affect Plants and Animals?

Imagine that a new road is being built. Huge, powerful machines rip up trees and clear away the soil. It's amazing how fast an area can be cleared. **Wait!** That's got to affect the environment, doesn't it?

How People Change Environments

Like all living things, people need habitats. People need food, shelter, and water. When people build their habitats, they change the environments of other living things.

Think about where we get food. Much of our food is grown on farms, like the one shown. Farmers need to change the land in order to produce food. Sometimes they need to remove trees or drain wetlands to make the land better for raising plants. They also might need to dig new paths for rivers and streams to bring in water for crops and farm animals.

In what ways has this farm changed the environment? ▼

People also cut down trees and dig up soil in order to build shelter. People changed the environment when they built the houses in the picture below.

Small towns, like this one, might grow larger. More people come, and they build more homes. Each time a house is built, trees might be cut down. More ground is dug up and more cement is put down.

Many people live far from farms. Think about how food gets from farms to stores in towns and cities. Food is carried on trucks, trains, and airplanes. People need to build roads, railroads, and airports so they can move the things they need from place to place.

This area has houses now. How did it look before the houses were built? And before that? ▼

Glossary

pollution
(pə lü′shən), anything harmful added to the air, water, or land

How Pollution Affects the Environment

Cities and the suburbs around them will be getting larger. There will be more cars, buses, trucks, and even airplanes. More people will need to get from place to place. More food and other supplies will need to be delivered.

Problems arise as more people travel from place to place. For example, cars and buses add harmful materials to the air. **Pollution** is anything harmful added to the air, water, or land. Pollution can harm living things and destroy their habitats.

Earth Science

People cause air pollution when they burn fuels in cars, trucks, homes, and factories. Gasoline, coal, and oil are some kinds of fuels. When fuels burn, they give off harmful materials. These materials get into the air.

Air pollution can be harmful to people. It also can harm plants and buildings. The picture shows one way to reduce air pollution. Can you think of other ways?

◀ Habitats changed when this road was built. Air pollution is bringing more changes. How can car pools help reduce air pollution?

Water pollution is another kind of pollution. Rain and melting snow can wash dirt and other materials into the water. Wastes from cities, factories, and farms can get into the water. Some of these wastes can be poisonous to the plants and animals living in the water. The beach shown in the picture looks clean. It is closed because the water is polluted.

Litter can also harm organisms that live in water habitats. For example, a sharp edge of a metal can might cut an animal. Plastic loops floating on water might hurt birds who are trying to catch fish. How can people prevent this kind of water pollution?

The garbage dump in the photo shows how litter and trash can pollute the land. Some materials in garbage might be harmful to living things. All trash should be put into special containers. From there, it will be taken to safe places.

Piles of garbage might spread diseases. They also might contain materials that harm the soil. ▼

This beach is dangerous for people. It would also be harmful to organisms that live in the water. ▼

Glossary

recycle (rē sī′kəl), to change something so it can be used again

How People Protect Habitats and Living Things

Groups of people all over the world are working to save certain plants and animals. Without their help, these organisms might become endangered or extinct. People can protect other living things by protecting their habitats. Read about some ways that people protect habitats and living things.

▲ Plant Trees

Trees are habitats for birds and other animals. Trees help the soil. They also add helpful gases to the air.

▲ Recycle

Here's something everyone can do to help habitats. Paper is made from trees. If people use less paper, then fewer trees might be cut down. We can also **recycle** *paper.*

◀ Protect Wetlands

Wetlands are areas of land that are sometimes or always filled with water. Swamps and marshes are kinds of wetlands. Plants grow in wetlands. These areas also are habitats for many insects, fish, and birds. Many wetlands have been drained and built into places for people to live and work. People are trying to protect other wetlands. They want these areas to become parks. Then the habitats will be protected.

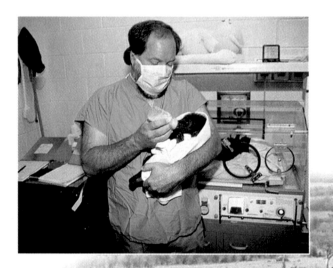

◄ Zoos

You might like to visit a zoo to look at the animals. This man is a veterinarian at a zoo. He is a doctor who takes care of animals. Other scientists use zoos to study animals. Zoos also protect endangered animals. Some zoos raise babies of endangered animals. When the babies are old enough to find food, zoo workers put them into their wild habitats.

▲ National Parks

National parks are places that are protected by laws. Parks like Masai Mara in Kenya protect plants and animals. No one is allowed to build houses in the park. People cannot hunt animals here. No one can collect plants or animals here either. None of the habitats in the park can be disturbed. Countries all over the world have national parks.

Lesson 3 Review

1. What are some ways that people change the habitats of other living things?

2. How might pollution affect living things?

3. What are some things people can do to protect habitats and living things?

4. **Predict**
   Suppose that someone wanted to build houses in a national park. Predict all the things that would happen to the park if the person could build there.

Cleaning Polluted Water

Process Skills

- observing
- inferring

Materials

- 2 cups of polluted water
- hand lens
- clock or timer
- one cup of tap water
- 1 L plastic bottle cut in half
- coffee filter
- masking tape
- marker
- clear plastic cup

Getting Ready

In this activity you can find out how to clean polluted water by filtering it. The water you filter will not be clean enough to drink. This activity will not remove all types of pollution.

Follow This Procedure

❶ Make a chart like the one shown. Use your chart to record your observations.

Type of water	Observations
Polluted water	
Polluted water after standing 5 minutes	
Tap water	
Filtered water	

❷ Use a hand lens to look at the cups of polluted water. Record your **observations.** Let the polluted water stand for at least five minutes. Observe the polluted water again. Record your observations.

❸ Look closely at the cup of tap water. Record your observations in your chart.

❹ Place the coffee filter in the top half of the plastic bottle. Place the top half of the bottle upside-down in the other half of the bottle (Photo A).

Photo A

⑤ Write *Filtered water* on a piece of masking tape. Put the label on an empty, clear plastic cup.

⑥ Slowly pour one of the cups of polluted water through the filter (Photo B). Do not pour any of the muddy material from the bottom of the cup.

⑦ Pour some of this filtered water into the labeled plastic cup. Compare and contrast the filtered water with the tap water and the polluted water. Record your observations.

> ⚠ **Safety Note** *Do not drink the filtered water.*

Interpret Your Results

1. What happened when you allowed the polluted water to stand for five minutes?

Photo B

2. How effective do you think the filter was in removing pollution? Explain.

3. Make an **inference.** Could the filtered water still contain some pollution that you can't see? Explain.

❓ Inquire Further

How can other materials that cause water pollution be removed? Develop a plan to answer this or other questions you may have.

Self-Assessment

- I followed instructions to make a water filter.
- I recorded my **observations**.
- I compared and contrasted the filtered water with the tap water and the polluted water.
- I evaluated how well the filter worked.
- I made an **inference** about pollution that cannot be filtered.

You will learn:

- about some foods that come from plants and animals.
- about some other products that come from plants and animals.

What Does the Environment Provide for People?

Do you ever go grocery shopping? There must be a hundred different kinds of cereal and canned goods! Notice all the fresh fruits and vegetables. Where does all this food come from?

People all over the world get their food in markets like this one. ▼

Foods from Plants and Animals

Many people do not shop for food in a supermarket. In some places, people get food in open markets. What kinds of foods can people get in this market in Mexico? Think about other kinds of foods they might be able to get in this market.

Many of the things in this market come from plants and animals. People all over the world depend on plants and animals for food. Meat, fish, eggs, and milk all come from animals. So why do we need plants? Remember, animals depend on plants.

People eat some plants just as they are. For example, you might pick an apple from a tree or pull a carrot from the ground. You would just wash it and eat it.

Other plants need to be changed before they can be used as food. Wheat is one plant that needs to be changed before people can use it for food. The farmer in the picture is cutting wheat plants. In many parts of the world, wheat is the most important food plant. People use more wheat than any other plant.

The wheat plant is a type of grain. Grains are plants whose seeds usually are ground into flour. Then the flour is used to make bread, cereal, pasta, tortillas, and many other foods. What is your favorite food made from wheat?

In other parts of the world, rice is the most important food plant. Rice grows in watery fields like the one shown. Rice grains have an outer coating that people do not eat. After the coating is removed, people cook and eat the grains. The outer coating of the grain is fed to the farm animals.

Wheat is one of the most important food plants in the world. ▼

Rice plants grow in watery fields. For people in many parts of the world, rice is their main food. ▶

Other Things from Plants and Animals

The environment gives people many things besides food. People also depend on plants and animals for shelter, clothing, medicines, and other products. As the pictures show, these products are all around us.

Shelter

This house is being built. You can see its frame. The wooden frame holds the house up. Wood comes from trees. The outside of the house might also be made of wood. What other things are made of wood? ▶

Medicines

Many medicines are made from plants. The seeds of some plants are used for some kinds of medicines. Leaves of other plants make other medicines. The bark of some trees is also used for medicines. ▶

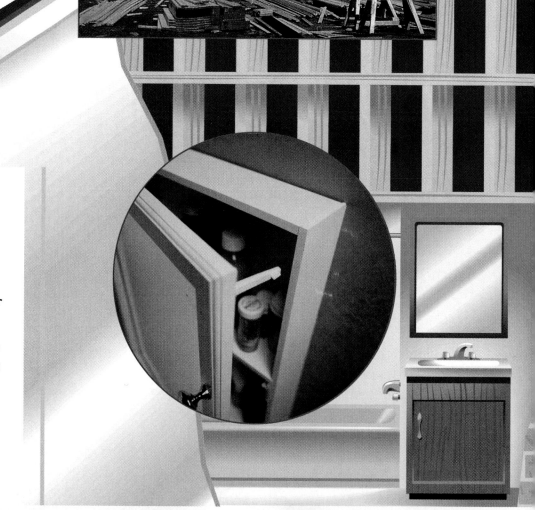

◄ Clothing and More

The boy's T-shirt is made of cotton cloth. Cotton plants have strong threads that are made into cloth. His sweater is made of wool cloth. Wool comes from the hair of sheep. Parts of his shoes are made of leather. People make leather from animal skins. The bottoms of his shoes are made of rubber. That's another product that can come from plants. Natural rubber is made from the sap of rubber trees.

◄ All Kinds of Paper

Look at all the ways we use paper. We cut down many trees to get all the paper we need. Now you see why it's important to use paper carefully and to recycle it.

Lesson 4 Review

1. What are two types of grain plants?

2. What are some things besides food that people get from plants and animals?

3. Predict
Suppose the wheat crop did not grow well for a long time. Predict some of the things that might happen.

Chapter 4 Review

Chapter Main Ideas

Lesson 1

• When a habitat changes, some organisms might no longer be able to live there.

• Too little or too much water can change a habitat.

Lesson 2

• Organisms that left when a habitat changed might return to it after a time.

• Some organisms may become endangered if their habitats change.

• Some organisms became extinct when their habitats changed or because of the actions of people.

Lesson 3

• People change environments in order to get food, shelter, and water.

• Pollution of air, water, and soil affects habitats.

• People can do many things to protect habitats and living things.

Lesson 4

• People depend on plants and animals for food.

• People also depend on plants and animals for shelter, clothing, medicines, and many other products.

Reviewing Science Words and Concepts

Write the letter of the word or phrase that best completes each sentence.

a. drought **d.** fossil

b. endangered organism **e.** pollution

f. recycle

c. extinct organism

1. Harmful material added to air, water, or soil causes ____.

2. A long period with not enough rain is a ____.

3. A living thing that may not be found on earth after a while is an ____.

4. A ____ is hardened material or marks left by an organism that lived long ago.

5. When people ____, they prepare materials so they can be used again.

6. A kind of organism that no longer lives on earth is an ____.

Explaining Science

Draw and label a diagram or write a list or paragraph to answer these questions.

1. What are some events that change a habitat?

2. How are endangered organisms different from extinct organisms?

3. Why do people change some habitats of other living things?

4. Why are plants and animals important for people?

Using Skills

1. Suppose that a city and the suburbs around it keep growing. **Predict** what will happen to the habitats in the area when new homes and roads are built.

2. Water is a very important part of your habitat. **Observe** all the ways that you use water. Write a list of your observations.

3. Some endangered animals live in national parks. Write a paragraph to **communicate** how national parks protect endangered organisms.

Critical Thinking

1. Some kinds of salmon hatch from eggs in rivers. When they grow older, the salmon follow the rivers to the sea, where they live. When the fish are ready to lay eggs, they swim up the same river. They lay their eggs near the place where they were hatched. Suppose that a dam is built across the river. Then the fish could not get over the dam. **Infer** what would happen to the salmon.

2. Think about why animals leave a habitat that has changed. Write a few sentences that explain your **main idea.**

3. Draw conclusions about how an endangered organism might become an extinct organism. Write a short paragraph that explains your conclusions.

4. Make a list of all the things you use that come from plants and animals. **Generalize** about how people help themselves when they protect plants and animals.

Unit A Review

Reviewing Words and Concepts

Choose at least three words from the Chapter 1 list below. Use the words to write a paragraph about how these concepts are related. Do the same for each of the other chapters.

Chapter 1
germinate
life cycle
petal
seed coat
seed leaf
seedling

Chapter 2
amphibian
embryo
mammal
nymph
pupa
tadpole

Chapter 3
adaptation
environment
food chain
habitat
predator
prey

Chapter 4
drought
endangered
  organism
extinct organism
fossil
pollution
recycle

Reviewing Main Ideas

Each of the statements below is false. Change the underlined word or words to make each statement true.

1. Green leaves use energy from sunlight to change water from the soil and <u>minerals</u> from the air into sugar and oxygen.

2. A <u>seedling</u> is all the stages in the life cycle of a living thing.

3. An <u>amphibian</u> is an animal before it is born or hatched.

4. Frogs belong to a group of animals called <u>mammals</u>.

5. A <u>nymph</u> is an animal that has a backbone and hair or fur.

6. An organism gets everything it needs from its <u>adaptation</u>.

7. Gills and fins are two <u>habitats</u> that help a fish live underwater.

8. A <u>consumer</u> is an organism that can make its own food.

9. A long period of dry weather without enough rain is a <u>flood</u>.

10. <u>Extinct organisms</u> are kinds of living things that are very few in number.

Interpreting Data

The following pictograph shows the number of trumpeter swans in different years. Use the pictograph to answer the questions below.

1. How many swans were alive in 1850? in 1950? in 1998?

2. What happened to the trumpeter swan population between 1850 and 1900? between 1900 and 1950? between 1950 and 1998?

3. Based on information in the pictograph, how might the trumpeter swan population continue to change?

Communicating Science

1. Draw and label the stages in the life cycle of a flowering plant. Write a sentence explaining what happens in each stage.

2. Explain the difference between how a nymph looks compared to an adult cockroach and how a larva looks compared to an adult butterfly.

3. Explain why all animals are consumers.

4. Explain how the building of a shopping center could affect habitats.

Applying Science

1. A friend is going to plant a small flower garden. He asks you for information about growing plants from seeds. Write a letter telling him what a seed needs to germinate and the importance of a plant's roots, stems, leaves, and flowers.

2. You are going to buy a hamster. Make a list of what the hamster will need in its new habitat.

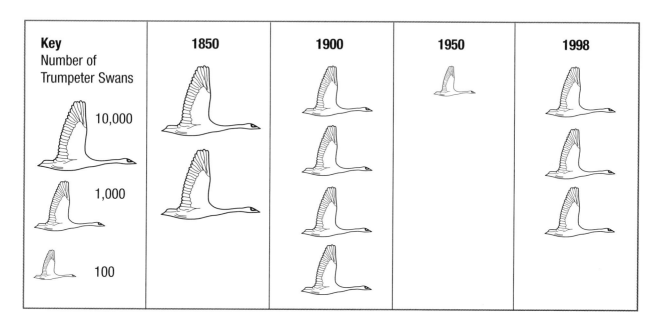

Key Number of Trumpeter Swans	1850	1900	1950	1998

Unit A
Performance Review

Habitat Day

Using what you learned in this unit, complete one or more of the following activities to be included in a Habitat Day event. The event will help students and visitors learn more about how organisms live in a habitat. You may work by yourself or in a group.

Geography

Look at a map of the world or a globe. Find deserts in different areas of the world. Choose one desert. Do research to find out about the temperature, rainfall, and organisms that live in that desert habitat. Make a poster that shows what you found out.

Environment

What if your town decided to clear wooded land to build a shopping center? How might this affect the food chain on the land? Suppose that you are a television reporter. Write a television news show that reports the plans and the changes that might take place.

Drama

Plan a puppet show in which you act out one day in the life of desert or forest plants and animals. Make puppets for the different plants and animals.

Music

Choose a popular tune. Use the tune to write a song about the life cycle of a plant or animal. Perform your song for the class. Use a sound recorder to record your song.

Diorama

Build a shoe-box diorama of a habitat. First, paint the background of your habitat. Then make models of organisms out of clay, papier-mâché, or construction paper.

Using Notes to Summarize

Taking good notes is an important study skill. Good notes help you get your thoughts in order before you write. You can use your notes to summarize or review what you have read. You can take notes in many different ways. For example, making lists, sketches, and charts are all good ways to take notes. Making a glossary is also a good way to take notes.

Take Notes

Review Chapter 1 to find out what each part of a plant does to help the plant live and grow and form seeds. Take notes while you review the chapter. Use one or more of the note-taking ideas discussed to organize your notes.

Write a Summary

Use the notes you took to write a two-paragraph summary about how different parts help the plant live and grow and form seeds. Your summary should describe the parts of a plant that help it get food, carbon dioxide, and water. It should also tell how flowering plants or cone-bearing plants form seeds. Make sure your summary covers all the basic points.

Remember to:

1. **Prewrite** Organize your thoughts before you write.

2. **Draft** Write your summary.

3. **Revise** Share your work and then make changes.

4. **Edit** Proofread for mistakes and fix them.

5. **Publish** Share your summary with your class.

Unit B
Physical Science

Science and Technology

In Your World!

Smother It!

If a fire breaks out, this tool helps save the day. Fire extinguishers contain liquid water, gases, or dry, powdery solids. In order to burn, fires need oxygen. The spray from the fire extinguisher covers the fire. Then oxygen cannot feed the flames. You'll learn more about solids, liquids, and gases in **Chapter 1 Matter and How It Changes.**

Going Up, Going Down!

With the push of a button, an elevator takes you to a different floor. You can see through the glass walls of some elevators. What you can't see are the weights, ropes, and pulleys that use forces to move the elevator up and down. You will learn more about making things move in **Chapter 2 Forces, Machines, and Work.**

Red Light "Reads" Your Groceries

Lasers are machines that produce very, very narrow light beams. Have you seen the red laser beams used in stores? The beam is formed when electricity is passed through gases. The beam is reflected off the bar code, and picked up by the scanner. The lines on a bar code tell what the item is, how much it costs, and so on. You'll read more about light and other forms of energy in **Chapter 3 Energy in Your World.**

You Could Be There!

You put on a headset. Suddenly you are zooming around a race track. Sounds from tiny speakers in the headset seem to come from all directions, just as in real life. Virtual reality games use light and sound to make it seem as if you are in the middle of all the action. You will learn more about sound in **Chapter 4 Sound.**

Make It! Break It!

You tear newspaper and soak it in paste. Then you shape the wet paper around a balloon. Add colored paper and you've made a piñata! You've changed some materials. Another change will happen when the piñata breaks.

Chapter 1
Matter and How It Changes

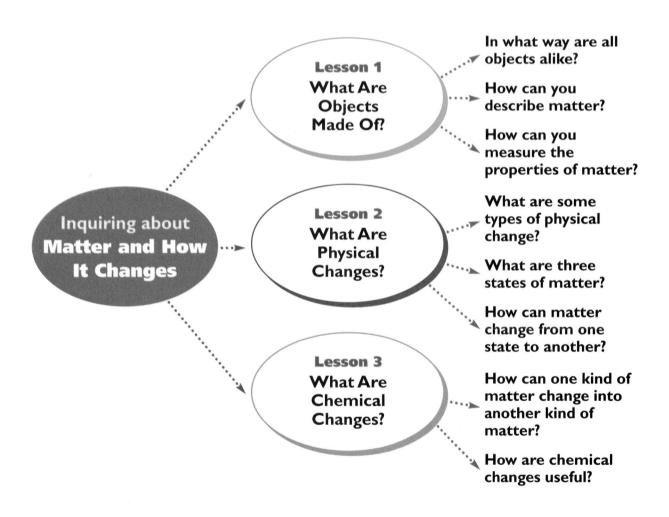

Inquiring about **Matter and How It Changes**

Lesson 1
What Are Objects Made Of?

In what way are all objects alike?

How can you describe matter?

How can you measure the properties of matter?

Lesson 2
What Are Physical Changes?

What are some types of physical change?

What are three states of matter?

How can matter change from one state to another?

Lesson 3
What Are Chemical Changes?

How can one kind of matter change into another kind of matter?

How are chemical changes useful?

Copy the chapter graphic organizer onto your own paper. This organizer shows you what the whole chapter is all about. As you read the lessons and do the activities, look for answers to the questions and write them on your organizer.

Exploring Properties of Matter

Process Skills

- observing
- making operational definitions
- communicating

Materials

- rock
- 2 empty plastic bags
- water in 2 plastic bags

Explore

1 Place the rock in a plastic bag. Seal the bag. Squeeze and press on the bag. Try to bend the rock to change its shape. Record your **observations.**

2 Observe the water in the bags. Gently squeeze and press on the bags. How does the shape of the water change? Record your observations.

3 Open the last plastic bag a small amount. Blow air into the small opening and seal it. Gently squeeze and press on the bag. How does the shape of the air space change? Open the bag and squeeze it. What happens to the air? Record your observations.

Reflect

1. The rock is a solid. The water is a liquid. The air contains gases. Use your observations to write an **operational definition** for a solid, a liquid, and a gas.

2. Communicate. Discuss your definitions with the class.

? Inquire Further

Which foods contain solids, liquids, and gases? Develop a plan to answer this or other questions you may have.

Measuring Volume: Metric Units

A **milliliter** (mL) and a **liter** (L) are units of volume in the metric system. Volume is the amount of space an object takes up or holds. The word *capacity* is sometimes used instead of volume. The capacity of a container is how much it can hold. You can estimate and compare measurement in liters and milliliters.

Math Vocabulary

milliliter
(mil′ə lē′tər), a metric unit of volume or capacity smaller than a liter

liter (lē′tər), a metric unit of volume or capacity equal to 1,000 mL

▲ *about 1 milliliter*

▲ *1 liter*

There are 1,000 milliliters in 1 liter.

Example 1
About how much soup does this soup spoon hold: 15 mL or 15 L?

Think: A liter bottle holds much more than a soup spoon. So, 15 mL is a better estimate.

Example 2
About how much water will the pail hold: 7 mL or 7 L?

Think: The pail is bigger than the liter bottle. The pail will hold several liters.

So, 7 L is a better estimate.

Talk About It!

Would you measure the amount of water in a full bathtub in liters or milliliters? Explain.

What's the Big Idea?

You will learn:

- about the way that objects are alike.
- about ways to describe matter.
- how to measure some properties of matter.

Glossary

matter, anything that takes up space and has weight

Lesson 1

What Are Objects Made Of?

Wow! Look at this mess! Did you ever think a closet could hold so many things? These objects are different, but they are all made of the same thing. How can that be?

How Objects Are Alike

Some things from the closet are small. Some are large. Some things look a lot heavier than others. Even so, every object takes up some space and has weight.

Look around the room. What kinds of objects do you see? You might see solid objects, such as desks. You might see liquids, such as water in a fish tank or in a glass. The air you breathe is a gas. You cannot see it, but it is all around you.

All the objects in the closet and in your classroom are alike in one way. Everything you see takes up space and has weight. Anything that takes up space and has weight is made of **matter**. Even the air that you cannot see is made of matter. The way that objects are alike is that they're all made of matter.

Everything here is alike in one way. ▶

An object can be made of different kinds of matter. Look at the ice skate. What kinds of matter are in it? People are made of several kinds of matter too. Describe some kinds of matter that make up people.

Lace is made of cotton.

Blade is made of metal.

Blade holder is made of plastic.

Boot is made of leather.

Glossary

property
(prop′ər tē),
something about an object—such as size, shape, color, or smell—that you can observe with one or more of your senses

Describing Matter

Suppose you have to tidy up the closet. Think about having to put those objects into groups or sets. The objects in each group would be alike in some way. They would be different from objects in other groups. The objects shown are grouped by how they are used. You also might group these objects by their color, size, or shape.

How would you describe these objects? You might say that some of them are hard and some are soft. Some things are solid and some are liquid. You would be describing the objects according to their properties. A **property** is something you can observe with one or more of your senses.

▲ What are some properties of these shoes?

▲ Different toys have different properties.

◀ Each ball has been given special properties because it is used to play a certain game.

B 10

Measuring the Properties of Matter

Length is a property of an object. It is one property that can be measured. Length tells what an object measures from end to end. Scientists measure length in meters or centimeters. A meter stick, ruler, and a tape measure are tools for measuring length.

Compare the length of the toy dinosaur with the length of the puzzle box. If you could use a meter stick to measure the length of each, you would know exactly how much shorter the puzzle box is than the dinosaur.

What property does the meter stick measure? ▶

▲ *How many centimeters long is this dinosaur model?*

You have seen how a meter stick is used to measure length. Other properties of matter can be measured too. Different tools are used for measuring other properties.

The amount of space an object takes up is its **volume.** A larger object has a greater volume than a smaller one. Volume is usually measured in liters or milliliters. The picture shows some tools for measuring volume.

Which container in the picture has the largest volume? What is the volume of milk that you drink with your lunch? How can you find out?

So Much Milk!
The jug, the measuring cup, and the graduated cylinder hold different volumes of matter. Volume and capacity are two ways to say how much a container can hold. ▼

You can also measure the amount of matter in an object. The amount of matter in an object is its **mass**. A heavy object has more mass than a light one. Mass usually is measured in grams. The mass of a paper clip is about one gram.

Mass and weight are not the same. An object's weight can be different in different places. An object's mass is always the same. For instance, astronauts in space seem to have no weight, but they have the same mass as they have on the earth.

A balance is a tool that is used for measuring mass. On a balance like the one shown, you put objects on the pans on both sides. When the arrow in the middle points to the center, the objects on both sides are in balance. That means they have the same mass. The objects on the balance in the picture have the same mass. Which object has the greater volume?

▲ **Measuring Mass**
The cereal box and the can of fruit have the same mass. The arrow on the balance is in the center and the arms are straight and flat.

Lesson 1 Review

1. How are all objects alike?

2. How can you describe matter?

3. How can you measure some properties of matter?

4. **Measure Volume: Metric Units**
   There are 1,000 milliliters (mL) in 1 liter (L). About 10 drops from a medicine dropper equals 1 mL. You can buy a bottle of water with a volume of 1 liter. Would you measure the volume of water in a large aquarium in milliliters or liters?

Estimating and Measuring Mass

Process Skills

- estimating and measuring
- collecting and interpreting data
- inferring

Materials

- pencil
- eraser
- paper clip
- piece of chalk
- gram cubes
- 2 plastic cups
- balance
- masking tape

Getting Ready

You can use your hands to estimate masses. Then use a balance to measure the actual mass of the objects.

Follow This Procedure

1 Make a chart like the one shown. Use your chart to record your data.

Object	Estimated mass	Measured mass
Pencil		
Eraser		
Paper clip		
Chalk		
Gram cube shape		
Gram cube shape taken apart		

2 Gather the objects you will measure. Make another object by joining 12 gram cubes together in any shape you choose.

3 Place one cup at each side of the balance. Use two small pieces of tape to attach each cup to the balance. Be sure the two cups are in balance.

4 Now **estimate** and **measure** the mass of different objects. Hold the pencil in one hand. Have a partner place one gram cube at a time in your other hand (Photo A). When you think the mass is the same in both hands, say "Stop."

5 Count the cubes in your hand. This is your estimate of the pencil's mass. **Collect data** by recording your estimate.

B 14

Photo A

⑥ Place the pencil in one of the cups on the balance. Add one gram cube at a time to the other cup until the cups balance (Photo B). Count the cubes in the cup. This is the pencil's mass. Collect data by recording your measurement.

⑦ Repeat steps 4-6 for the eraser, paper clip, chalk, and gram cube shape.

Photo B

⑧ Take the gram cube shape apart and repeat steps 4-6 one last time.

Interpret Your Results

1. Compare your estimated masses with the measured masses. How close were your estimates?

2. How many grams were needed to balance the gram cube shape when it was whole? when it was taken apart? What **inference** can you make about the mass of an object when it is whole and when it is taken apart?

Inquire Further

How could practice affect your ability to estimate masses? Develop a plan to answer this or other questions you may have.

Self-Assessment

- I followed instructions to use a balance.
- I followed instructions to **estimate** and **measure** the mass of different objects.
- I **collected** and recorded my **data.**
- I compared the estimated mass of different objects with the actual mass of the objects.
- I made an **inference** about the mass of an object when it is whole and when it is taken apart.

You will learn:

- about the types of physical change.
- about three states of matter.
- how matter can change from one state to another.

Glossary

physical (fiz′ə kəl) **change,** a change in the way matter looks, but the kind of matter remains the same

You might cut melons into small pieces. What kind of change have you made? ▶

Lesson 2

What Are Physical Changes?

You take some crayons and draw a picture on a piece of paper. You might not realize that you've made a change. You've actually made a physical change to the paper.

Types of Physical Change

Every time you draw a picture, cut some paper, or pour juice into a cup, you are changing matter. The matter you have changed looks different, but it is still the same kind of matter. You have changed only some of its properties. The paper is still paper and the juice is still juice.

When you change the way matter looks, you make a **physical change**. Matter can go through many kinds of physical change.

Suppose you dig into a watermelon and remove pieces, as the girl is doing. Then you might cut up apples and bananas. You would have made physical changes to all the fruits.

You might put all those pieces of fruit into a bowl. Then you take a large spoon and mix them all together. You've made a fruit salad! You've also made a mixture. A **mixture** is made of different kinds of matter that are placed together. Each kind of matter can be easily separated from the others. What are some other mixtures?

You might make juice from some of the fruit. This is a physical change. You could change some of the juice by freezing it into a bar, like the one that the boy in the picture is trying to eat. When the juice froze, it went through another physical change. In the warm air, the juice bar is changing again.

Glossary

mixture
(miks′chər), two or more kinds of matter that are placed together but can be easily separated

Could you separate your favorite fruit from this mixture? ▼

◀ *Sometimes physical changes happen quickly. You might want them to occur more slowly.*

Glossary

states of matter, the three forms of matter—solid, liquid, and gas

solid (sol′id), a state in which matter has a definite shape and volume

liquid (lik′wid), a state in which matter has a definite volume but no shape of its own

gas, a state in which matter has no definite shape or volume

States of Matter

Remember what you learned about the properties of matter. You learned that matter could be a solid, liquid, or gas. Look at all the matter in the aquarium. Each kind of matter has one of three forms—solid, liquid, or gas. These forms are sometimes called the **states of matter.**

Matter that is in the **solid** state has a definite shape and volume. Find some solid objects in the picture.

Everything in the fish tank is in water, which is a liquid. Matter in the **liquid** state has a definite volume, but has no shape of its own. A liquid takes the shape of its container.

Notice the air bubbles in the water. Air is a gas. In the **gas** state, matter has no definite volume or shape. Like a liquid, a gas takes the shape of its container. If a container is open or breaks, some gas comes out and mixes with air in the room. The air bubbles in the tank are going up. When they reach the top of the water, the bubbles break. The air comes out of the water. It mixes with the air in the room.

What forms of matter do you see? ▼

Changes of State

Matter can change from one state to another state. When matter changes state, it goes through a physical change.

Water changes its state when you heat it or make it very cold. When water is cooled to 0°C, it changes from liquid to solid. You know what happens when ice gets warm.

You can see water as a liquid and as a solid. You cannot see water as a gas. However, water does exist as a gas called **water vapor.** You probably have noticed a puddle that disappeared after a while. The water in it became water vapor. When a liquid **evaporates,** it turns into a gas. When water boils, it evaporates quickly.

Notice the water drops outside the cold glass of juice. Where did that water come from? You didn't see it happen, but water vapor in the air touched the cold glass. Then the gas turned into a liquid. When a gas **condenses,** it changes into a liquid.

Glossary

Glossary

water vapor
(vā′pər), water in a gas state

evaporates
(i vap′ə rāts′), changes from a liquid state to a gas state

condenses
(kən dens′iz), changes from a gas state to a liquid state

▲ When water vapor in the air touches the cold glass, it turns into the water drops that you see on the outside of the glass.

▲ When water boils, it changes quickly into water vapor. Water vapor is in the bubbles in the water. You cannot see water vapor.

Lesson 2 Review

1. What happens in a physical change?

2. What are the differences among the three states of matter?

3. How does water change state?

4. **Measure Volume: Metric Units**
   You breathe about 6 liters of air every minute. How many liters of air do you breathe in 5 minutes?

Changing States of Matter

Process Skills

- observing
- estimating and measuring
- inferring

Materials

- large ice cube
- 2 plastic jars with lids
- measuring cup
- water
- food coloring
- paper towel

Getting Ready

Water can be a solid, a liquid, and even an invisible gas. In this activity you can observe the properties of solid and liquid water, and make an inference about water as a gas.

Follow This Procedure

1 Make a chart like the one shown. Use your chart to record your observations.

Drawings of observations

	Jar upright	Jar at angle	Jar on side
Ice			
Water			

2 Place an ice cube in a jar. Put the lid on the jar. Place the jar upright on your desk. Make a drawing of the ice in the jar.

3 How will the shape of the ice change if the jar is held at an angle? Have a partner hold the jar at an angle (Photo A). Make a drawing of your **observations** of the jar and the shape of the ice.

4 Repeat step 3, but this time turn the jar completely on its side.

Photo A

Photo B

5 **Measure** 100 mL of water and pour it into the other jar. Put two drops of food coloring in the water. Put the lid tightly on the jar. Place the jar upright on your desk. Make a drawing of your observations of the jar and the shape of the water.

6 Repeat steps 3 and 4 with the jar of water instead of the jar with ice (Photo B).

⚠️ *Safety Note Clean up all spills immediately.*

7 Place the jar upright. Place the ice cube in the jar of water and replace the lid. Look closely at the outside of the jar for about three minutes. You should see moisture forming on the outside of the jar. Wipe some of the moisture off with a paper towel. Is there any coloring in the moisture?

Self-Monitoring
Do I have questions to ask before I continue?

Interpret Your Results

1. How are the shapes of solid and liquid water different from each other?

2. Make an **inference.** Where did the moisture on the outside of the jar come from? Explain.

**Inquire Further**

If you left an uncovered jar of water out for several days, where would the water go? Develop a plan to answer this or other questions you may have.

Self-Assessment

- I followed instructions to observe solid ice and liquid water.
- I **measured** the amount of water in the jar.
- I recorded my **observations.**
- I described differences between water as a solid and a liquid.
- I made an **inference** about where the moisture on the outside of the jar came from.

You will learn:

- how one kind of matter can change into a different kind.
- about ways that chemical changes are useful.

Glossary

chemical (kem⁄ə kəl) **change,** a change that causes one kind of matter to become a different kind of matter

This truck used to be shiny. Rust and shiny steel are different kinds of matter. ▼

Lesson 3

What Are Chemical Changes?

Yum! The smell of baking bread is coming from the bakery. So what does baking bread have to do with science? Actually, it has a lot to do with science. Baking bread is an example of a chemical change.

Changing into Different Materials

Suppose a wood block is cut into a different shape. What type of change takes place? Then suppose the wood block is put into a fire. As the fire burns, you feel heat and see smoke. When the fire burns out, ashes remain. The wood has changed into different kinds of matter. It changed into gases and ashes. They are not at all like the wood block.

A **chemical change** takes place when one kind of matter changes into a different kind of matter. This toy truck was once made of firm, shiny steel. Someone left it outside in the rain. Over time, the steel combined with water and air and turned into rust. Rusting is a chemical change.

A chemical change takes place when fireworks explode. Energy is given off during the chemical change.

Think about a melting juice bar. If you let it melt into a bowl, you could freeze it again. What about the rust on the truck? Can it ever again become shiny steel? Probably not. Materials that have gone through a chemical change usually cannot be changed back to the first kind of matter.

The pictures show two other kinds of chemical change. What other ways can you think of in which matter changes into a different form?

When the apple was just bitten, it was a light color. After the apple sat in air for a while, it turned soft and brown. The part where the bite was taken went through a chemical change in the air. How does the apple's skin help the apple? ▼

▲ *The batter for banana bread is a mixture. It has many things mixed together.*

▲ *The oven's heat helps the chemical change happen.*

▲ *How is the batter different from the bread?*

Using Chemical Changes

Human Body

Chemical changes are very useful. They happen in kitchens all the time. The students in the picture are mixing batter for banana bread. Compare the baked bread with the batter.

While the bread was baking, the batter changed into different kinds of matter. A gas formed during this chemical change. Bubbles of the gas made the batter spread out. Imagine the delicious smell in the kitchen! It comes from the matter that formed during the chemical change.

Suppose you could eat a slice of that bread. Then other chemical changes would happen in your body. The bread would change into different forms of matter that your body can use.

Many things that we use every day are made by chemical changes. Plastics, some clothing materials, and some medicines are a few of the things that are made by chemical changes.

Lesson 3 Review

1. When one kind of matter changes into a different kind of matter, what kind of change has taken place?

2. What are some ways that we use chemical changes?

3. **Measure Volume: Metric Units** Suppose you mix one spoonful of baking soda into batter for banana bread. Would you use 15 mL or 15 L of baking soda?

Experimenting with a Chemical Change

Materials

- safety goggles
- water (room temperature)
- measuring cup
- 3 plastic cups
- 3 antacid tablets
- clock with a second hand
- resealable plastic bag
- metal spoon

Process Skills

- formulating questions and hypotheses
- identifying and controlling variables
- experimenting
- estimating and measuring
- collecting and interpreting data
- communicating

State the Problem

How does changing the shape of an antacid tablet affect how fast it breaks down in water?

Formulate Your Hypothesis

If you break apart or crush an antacid tablet, will it break down in water faster than, slower than, or at the same rate as a whole tablet? Write your **hypothesis.**

Identify and Control the Variables

The shape of the tablet is the **variable** you can change. In Trial 1, use one whole tablet in one cup of water. In Trial 2, break a tablet into four pieces before putting it in water. In Trial 3, crush a tablet before putting it in water. Keep the amount and temperature of the water the same for each trial.

Test Your Hypothesis

Follow these steps to perform an **experiment.**

1 Make a chart like the one on page B27. Use your chart to record your data.

Continued →

Photo A

Photo B

2 Put on your safety goggles.

3 Put 150 mL of water in each plastic cup.

⚠ **Safety Note** *Wipe up any spills immediately.*

4 For Trial 1, drop one antacid tablet into one cup of water. Do not stir the water (Photo A). Use the clock to **measure** how long it takes for the tablet to break down completely. **Collect** and record the **data** in your chart.

⚠ **Safety Note** *Do not drink the liquid or eat any of the antacid.*

5 For Trial 2, break the second tablet into four pieces. Add the pieces all at once to the second cup of water (Photo B). Use the clock to measure how long it takes for the tablet to break down completely. Collect and record the data in your chart.

6 For Trial 3, put the third antacid tablet into the plastic bag and seal the bag. Use the metal spoon to crush the tablet. Pour the crushed tablet all at once from the plastic bag into the third cup (Photo C). Use the clock to measure how long it takes for the tablet to break down completely. Collect and record the data in your chart.

Photo C

Collect Your Data

Trial	Tablet shape	Time to break down
1		
2		
3		

Interpret Your Data

Label a piece of grid paper as shown. Use the data from your chart to make a bar graph on your grid paper. Study your graph. Describe any patterns you see in the time it took for the tablets to break down.

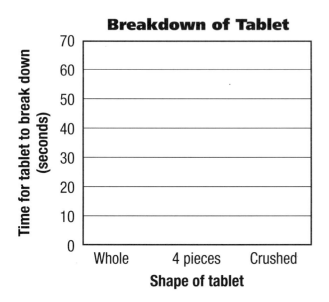

Breakdown of Tablet

Time for tablet to break down (seconds) — 0, 10, 20, 30, 40, 50, 60, 70

Shape of tablet — Whole, 4 pieces, Crushed

State Your Conclusion

How do your results compare with your hypothesis? **Communicate** your conclusion. Explain how changing the shape of the tablet affected the time it took to break down.

Inquire Further

If you increase the temperature of the water, will a tablet break down faster? Can you think of other ways to make the tablet break down faster? Develop a plan to answer these or other questions you may have.

Self-Assessment

- I made a **hypothesis** about an antacid tablet breaking down in water.
- I **identified** and **controlled variables**, and I followed instructions to perform an **experiment** with antacid tablets.
- I **measured** the time it took for the tablets to break down.
- I **collected** and **interpreted data** by making a chart and a bar graph.
- I **communicated** by stating my conclusion.

Chapter 1 Review

Chapter Main Ideas

Lesson 1

• All objects are alike because they are made of matter.

• Objects may be described by their properties.

• Length, mass, and volume are three properties of matter which can be measured.

Lesson 2

• A physical change is a change in the way matter looks.

• Matter exists in the solid state, liquid state, and gas state.

• When matter changes from one state to another, a physical change takes place.

Lesson 3

• Matter goes through a chemical change when one kind of matter changes into another kind of matter.

• Chemical changes are useful for making new kinds of matter.

Reviewing Science Words and Concepts

Write the letter of the word or phrase that best completes each sentence.

a. chemical change

b. condenses

c. evaporates

d. gas

e. liquid

f. mass

g. matter

h. mixture

i. physical change

j. property

k. solid

l. states of matter

m. volume

n. water vapor

1. In a ___, you change some properties of matter but the kind of matter remains the same.

2. Solid, liquid, and gas are ___.

3. Two or more kinds of matter that are put together but can be separated make up a ___.

4. You can observe a ___ of an object with your senses.

5. In a ___, matter has a definite shape and volume.

6. Water in a gas state is ___.

7. When water changes from a liquid to a gas, it ___.

8. Anything that takes up space and has weight is ___.

9. The amount of space an object takes up is its ___.

10. In a ___, matter becomes a different kind of matter.

11. In a ___, matter has a definite volume but no shape of its own.

12. An object's ___ is the measure of how much matter it contains.

13. Matter in the ___ state has no definite shape or volume.

14. When matter changes from a gas state to a liquid state, it ___.

Explaining Science

Draw and label a diagram or write a paragraph to answer these questions.

1. What are some properties of matter that you can measure?

2. How can liquid water change into the solid state and into the gas state?

3. What are some chemical changes that you have seen?

Using Skills

1. Suppose you put 250 mL of water into two cups. You put the cups in different places. After three hours, you **measure the volumes.** You **collect data** that you put in this chart. **Interpret** your **data** about warmth and evaporation.

Cup	Place	Volume
A	desk	230 mL
B	warm window	200 mL

2. Find several containers in your classroom. **Observe** the volumes in milliliters listed on the containers.

Critical Thinking

1. Sara is going to bake a banana bread. She has the batter mixed in a bowl. If she puts all the batter into one pan, it will come to the top of the pan. She could fill two pans half way. **Communicate** to Sara what you think she should do.

2. When sugar is heated for a long time, a solid black material forms. **Infer** what kind of change has taken place.

Jump!

To jump rope, you push and pull the rope around you. At the same time, you move up and down. Did you know that while you are having fun jumping rope, you are also doing work?

Chapter 2
Forces, Machines, and Work

Inquiring about Forces, Machines, and Work

Lesson 1
What Makes Things Move?

How does force make an object move?

How does friction affect a moving object?

Lesson 2
What Are Gravity and Magnetism?

How does the force of gravity affect objects?

What is magnetism?

Lesson 3
How Do Simple Machines Help You Do Work?

What must happen for work to be done?

How do levers, inclined planes, and screws make work easier?

How do a wedge, a pulley, and a wheel and axle help a person do work?

How do some animals use body parts as simple machines?

Copy the chapter graphic organizer onto your own paper. This organizer shows you what the whole chapter is all about. As you read the lessons and do the activities, look for answers to the questions and write them on your organizer.

Exploring the Motion of a Ball

Process Skills

Process Skills

- estimating and measuring
- communicating

Materials

- half-meter stick
- wall
- 2 short strips of masking tape
- 1 long strip of masking tape
- plastic-foam ball

Explore

1 Use the two short strips of tape to form an X shape about one meter from a wall. Attach the long strip of tape to the floor about one meter from the X shape as shown.

2 Place the ball on the center of the long strip of tape. Try to push the ball just enough to make it stop on or near the center of the X.

3 **Estimate** and record how close the ball is to the center of the X. **Measure** and record the distance from the ball to the center of the X.

4 Now repeat steps 2 and 3, but this time push the ball so it bounces off the wall and rolls onto or near the center of the X.

Reflect

Describe how you changed your push when you bounced the ball off the wall. **Communicate.** Discuss your descriptions with the class.

Inquire Further

How would the motion of the ball change if you moved the tape further from the wall? Develop a plan to answer this or other questions you may have.

Identifying Cause and Effect

A **cause** makes something happen. An **effect** is the result. Think about how you caused the ball to move in the activity *Exploring the Motion of a Ball.* Ask yourself what effect your actions had on the motion of the ball. In Lesson 1, *What Makes Things Move?,* you will learn more about cause and effect. As you read the lesson, notice the different ways to change how objects move. Look for the cause and effect of each change.

Example

Suppose you are playing in a baseball game. The pitcher throws the ball. You hit the ball. What are the cause and effect of these actions?

One way to better understand cause and effect is to make a chart. On a piece of paper, make a chart like the one below. Ask yourself what might have caused each effect to happen. Write the cause for each effect in your chart.

Cause	Effect
	The baseball changes direction.
	Your bicycle moves faster.
	The door to a room opens.
	The lights in the room go on.

Talk About It!

1. What is a cause? What is an effect?

2. What caused the ball to stop where it did in the activity *Exploring the Motion of a Ball?*

Reading Vocabulary

cause (kôz), a person, thing, or event that makes something happen

effect (ə fekt′), whatever is produced by a cause; a result

▼ *Have you ever wondered what might cause things to move?*

What's the Big Idea?

You will learn:
- how force makes an object move.
- how friction affects a moving object.

What Makes Things Move?

The sky is bright and clear. What a great day for a bicycle ride! You pull the door open and run outside. You climb on your bicycle and push down on the pedals. Away you go!

How Force Moves Objects

The door and the bicycle both needed a force to make them move. A **force** is a push or a pull. You use forces in many things you do during the day. You apply forces to pull open a drawer and to push it closed. You apply forces to zip up your coat and to sharpen your pencil. You are applying force any time you push or pull an object. Is the girl in the picture applying a pulling force or a pushing force to open the door?

◀ A force is needed to start an object moving. The girl is applying force to pull the door open. Without a force, the door will stay closed.

A force changes the way an object moves. An object will move in the direction it is pushed or pulled. If you change the direction of the force, the object will move in a different direction. Suppose someone had just kicked the ball to the girl in the picture. She kicks the ball back. She has changed the direction of the force that moves the ball. The ball will change direction.

An object moves faster when you apply more force. A bicycle will move faster when the rider pedals faster.

You need to apply more force to move heavier objects than lighter ones. Suppose you move a book. Your friend moves a pencil. Who is using more force?

◄ The girl is using force to change the direction the ball is moving.

Glossary

friction (frik′shən), the force caused by two objects rubbing together that slows down or stops moving objects

▲ Heat

Rub your hands together. How do they feel? When objects rub together, they become warm. This is because friction also causes heat.

How Friction Affects Objects

Suppose you apply the hand brakes to your moving bicycle. The brakes of the bicycle rub against the rim of the wheels. The wheels of the bicycle rub against the ground. You have used friction to slow down and stop your bicycle. **Friction** is a force that makes moving objects slow down or stop. Friction occurs between two objects that are rubbing against each other. Notice what is causing friction in the pictures on these pages.

When rough objects rub together there is more friction. When smooth objects rub together there is less friction. A sidewalk is rougher than a patch of ice. There is more friction between the bottom of your shoe and the rough sidewalk than between your shoe and the smooth ice. So it is easier to walk on a sidewalk than a patch of ice.

Slowing Down

The sled goes faster and faster as it speeds downhill. However, once the sled is on a level surface, friction produced between the bottom of the sled and the snow will make the sled slow down and stop. ▶

▲ Stopping
The girl is using force to move the golf ball across the ground. The ball is rubbing against the surface of the ground. Friction causes the ball to stop before it falls in the hole.

Lesson 1 Review

1. What is a force?

2. What does friction do to a moving object?

3. Cause and Effect
What are two ways force can cause an object to change the way it moves?

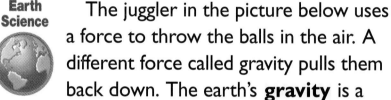**Lesson 2**

What Are Gravity and Magnetism?

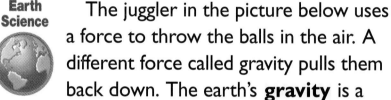**You will learn:**
- how the force of gravity affects objects.
- about magnetism.

Glossary

gravity (grav′ə tē), the force that pulls objects toward the center of the earth

You throw a ball in the air. You think you threw so hard the ball might be to the moon by now. A few steps later, you find the ball on the ground. Why didn't the ball keep moving in the direction you threw it?

The Force of Gravity

Earth Science

The juggler in the picture below uses a force to throw the balls in the air. A different force called gravity pulls them back down. The earth's **gravity** is a force that pulls objects toward the center of the earth. The pull of gravity keeps you on the ground. It keeps your desk on the floor and your papers on your desk. If you let go of your pencil, gravity would pull it toward the ground. Without gravity, anything that is loose would float away.

◀ *The juggler throws the balls into the air. Gravity pulls them back down toward the earth.*

B 38

Look at the pictures on this page. When leaves fall off a tree, the force of gravity pulls them to the ground. The tumbler uses force to jump in the air, but knows gravity will always bring him back down to the floor. Think of some other examples of gravity you have seen.

The pull of gravity gives you weight. The weight of an object tells how much gravity is pulling on it. Suppose you had no weight. You would be weightless. You would float around like the people who travel in the space shuttle.

The leaves from this tree would continue to float in the air if gravity did not pull them to the ground. ▶

▲ *A pushing force sends the tumbler into the air. The force of gravity brings him to the ground.*

Unlike poles pull together to hold each part of the toy train in place. ▶

Magnets and Magnetism

Magnets pull objects that are made of certain metals, such as iron. The force that magnets produce is called **magnetism**. Magnetism can be a pushing or pulling force. Look at the picture below. Notice how some objects stick to the ends of the bar magnets. The ends of a bar magnet are called **poles**. The objects that are sticking to the poles of the bar magnets are made of iron or have iron in them. The objects that are not sticking to the poles of the bar magnets do not have iron in them.

Glossary

magnetism (mag/nə tiz/əm), the force that causes magnets to pull on objects that are made of certain metals, such as iron

pole (pōl), a place on a magnet where magnetism is strongest

Magnets have two poles. Magnetism is strongest at a magnet's poles. ▼

Objects stick to a magnet's poles because that is where magnetism is strongest. The north pole is at one end of a bar magnet. It is usually marked N. The south pole is opposite from the north pole. The south pole is usually marked S. When hanging freely on the end of a string, the north pole of a magnet will point north. The south pole will point south. North and south poles are opposite, or unlike, poles. Unlike poles attract, or pull together. Like poles repel, or push away from, each other. Why do the pieces of the train in the picture above pull together? Why would the two poles in the picture to the right push apart?

▲ *If you try to put two like poles together, they repel, or push away from, each other. The push gets stronger as the like poles of a magnet get closer together.*

Lesson 2 Review

1. What is gravity?

2. What is magnetism?

3. **Cause and Effect**
   What force causes the wheels of your bike to stay on the ground and you to stay on the seat of your bike?

Investigating Magnetic Force

Process Skills

- observing
- inferring

Materials

- thread
- paper clip
- modeling clay
- donut magnet
- paper
- aluminum foil
- plastic sheet
- steel spoon

Getting Ready

You can find out how magnetic force acts by making a paper clip seem to float in midair.

Read the list of materials you are going to use in this activity again. Think of which materials might have iron or steel in them.

Follow This Procedure

1 Make a chart like the one shown. Use your chart to record your observations.

	Observations
Paper clip very close to magnet	
Paper clip farther from magnet	
Paper	
Aluminum foil	
Piece of plastic	
Metal spoon	

2 Tie the thread to one end of the paper clip (Photo A).

3 Press the clay onto your desk or other solid surface. Press the magnet into the clay so it is held in place. Put the end of the paper clip that is not tied to the thread on the magnet as shown (Photo B).

4 Pull slowly on the thread until the paper clip is pulled away, but still very close to the magnet. Hold the thread still. What happens? Record your **observations.**

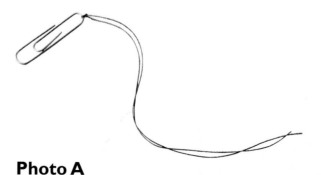

Photo A

Photo B

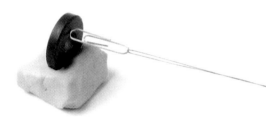

⑤ Slowly continue to pull the paper clip away from the magnet. What happens? Record your observations.

⑥ Put the paper clip back on the magnet. Pull slowly on the thread until the paper clip is held in midair by the magnet. Hold the thread very still. Have a partner carefully place a piece of paper between the paper clip and the magnet. Record your observations.

⑦ Repeat step 6 three more times using a piece of foil, a piece of plastic, and a metal spoon instead of the piece of paper. Record your observations.

Interpret Your Results

1. Think about what happened when you continued to pull the paper clip away from the magnet. What can you **infer** about a magnet's force as you move the paper clip farther away?

2. Look at your chart. Think about how each type of material affected the magnetic force. What can you infer about magnetic force and different materials?

 Inquire Further

What would happen if you increased the thickness of the materials between the magnet and paper clip? Develop a plan to answer this or other questions you may have.

Self-Assessment

- I followed instructions to **observe** how magnetic force acts.
- I observed how different materials affected the magnet and paper clip.
- I recorded my observations.
- I made an **inference** about magnetic force over distance.
- I made inferences about how different materials affect magnetic force.

What's the Big Idea?

You will learn:

- how work is done.
- about levers, inclined planes, and screws.
- about a wedge, a pulley, and a wheel and axle.
- how some animals use body parts as simple machines.

How Do Simple Machines Help You Do Work?

It is time to do your homework. You push a chair up to the table, pull your textbook toward you, and pick up a pencil. Guess what? You did work before you even started your homework!

Work

Work is done whenever a force makes an object move. You do work when an object moves in the same direction you push or pull it. You did work when you moved the chair, your textbook, and your pencil. The boy in the picture did work when he threw a plastic bottle into the recycle bin.

Glossary

work (wėrk), something done whenever a force moves an object through a distance

◄ *The boy did work because he was able to make the bottle move.*

The amount of work you do depends on how much force you apply to move an object. Suppose the girl below is bringing one puppy to show you. She did work when she picked up the puppy. Now imagine she is bringing two puppies to show you. The girl is doing more work. She used more force to lift two puppies and carry them the same distance.

The amount of work you do also depends on how far you move an object. You do more work when you use the same amount of force to move an object a greater distance. The girl did work when she lifted the puppy just above the floor. She did more work when she raised the puppy higher as she stood up.

If an object does not move, no work is done. The boy in the picture to the right is pushing as hard as he can against the wall. However, he is not doing work because the wall does not move.

▲ *This boy is not doing work because he cannot make the wall move.*

The girl did work by using force to lift the puppy. ▶

Glossary

simple machine
(sim′pəl mə shēn′),
one of six kinds of
tools with few or no
moving parts that
make work easier

lever (lev′ər), a
simple machine made
of a bar or board that
is supported
underneath at the
fulcrum

fulcrum (ful′krəm),
the point on which a
lever is supported and
moves

Lever, Inclined Plane, and Screw

A **simple machine** is a tool that has few or no moving parts. Simple machines help make work easier. There are six kinds of simple machines.

A **lever** is a simple machine made of a bar or board. The lever moves back and forth on a point. The point is called a **fulcrum.** The fulcrum supports the lever. What is being used as the fulcrum in the picture below?

Levers help you move objects. Notice how the person is pushing down on one end of the lever to make the object on the other end move. An object can be lifted more easily if you move the fulcrum closer to the object. A seesaw is an example of a lever.

Pushing down on one end of the lever makes it easier to move the object on the other end. ▶

An **inclined plane** is a simple machine with a flat surface. One end of an inclined plane is higher than the other end. Inclined planes help people move objects to a higher or lower place. The woman in the picture below is pushing the wheelchair up a ramp. A ramp is an inclined plane. It is easier to move the wheelchair using an inclined plane. Suppose the ramp was steeper. The woman would need more force to move the wheelchair. A path going up a hill and a slide are other examples of an inclined plane.

A **screw** is an inclined plane wrapped around a rod. The edge of the paper wrapped around the pencil in the picture is like the ridges on a screw. The ridges of a screw are the inclined plane. A screw is used to hold objects together.

Glossary

inclined plane
(in klīnd′ plān), a simple machine that is a flat surface with one end higher than the other

screw (skrü), a simple machine used to hold objects together

◄ *The edge of the paper is like the ridges on a screw.*

A force moves an object up or down an inclined plane. Suppose the woman in the picture had a choice of pushing the wheelchair up a ramp or pulling it up stairs. Which would make her work easier? Why? ▶

▲ Screws can hold objects together very tightly. A bolt is an example of a screw.

Glossary

pulley (pul′ē), a simple machine made of a wheel and a rope

wheel and axle (wēl and ak′səl), a simple machine that has a center rod attached to a wheel

gear (gir), a wheel with jagged edges like teeth

wedge (wej), a simple machine used to cut or split an object

Wedge, Pulley, and Wheel and Axle

You know that a simple machine has few or no moving parts. Every day people use simple machines to help make their work easier. Remember that there are six kinds of simple machines. You have already learned about three of them—the lever, the inclined plane, and the screw. Other simple machines include the wedge, the wheel and axle, and the pulley. Find examples of the wedge, wheel and axle, and pulley on these pages.

Pulley

*A **pulley** is a rope or chain wrapped around the edge of a wheel. Pulleys move an object up, down, or sideways. They help move a heavy object. Pulleys also help move an object to a place that is hard to reach. You pull down on the rope to raise the flag. The wheel at the top of the pole turns when the rope moves. The flag moves to the top of the flagpole. You use a pulley to open and close curtains and to raise and lower blinds.* ▶

Wheel and Axle

A **wheel and axle** is made of a rod, or axle, attached to the center of a wheel. The wheel turns around the axle. A wheel and axle moves or turns an object. The wheels on in-line skates or a bicycle help you move. The wheels on a wheelbarrow help you move objects. A doorknob is a wheel and axle. ▶

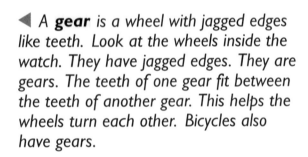

◀ A **gear** is a wheel with jagged edges like teeth. Look at the wheels inside the watch. They have jagged edges. They are gears. The teeth of one gear fit between the teeth of another gear. This helps the wheels turn each other. Bicycles also have gears.

Wedge

A **wedge** is made of two inclined planes that come together. Wedges are used to push or split objects apart. The bottom part of the ax in the picture is a wedge. It is used to split the piece of wood into two pieces. A knife is used as a wedge to cut a piece of bread from a loaf. A wedge under a door will keep the door open. ▶

A Rabbit's Claws

To make a burrow in the ground, a rabbit uses its claws as wedges to loosen soil. It then uses its claws and hind feet as shovels, or levers, to lift the dirt out. People also use shovels to lift dirt. ▼

Life Science

Some animals use parts of their body the same way people use simple machines. A burying beetle uses its front legs like a shovel or lever to dig into the ground. Carpenter ants have strong jaws that they use to chisel their way through wood. A chisel is a wedge. Look at the pictures on these pages. How do these animals use body parts as simple machines?

◄ A Bird's Beak

Some birds use their beaks like a pair of tongs to scoop up fish from the water or to pluck berries for food. Notice the tongs in the picture. Tongs are two levers fastened together at one end. You may have seen a pair of tongs in the kitchen. Most birds also use their beaks to pick up and carry nest-building materials.

A Beaver's Teeth

Beavers have long front teeth with sharp edges. The chisel in the picture also has sharp edges. People use a chisel as a wedge to cut wood. Beavers use their teeth as a chisel, or wedge, to cut down trees. ▶

Lesson 3 Review

1. How can you tell if you are doing work?

2. How do levers, inclined planes, and screws help people do work?

3. How do a wedge, a pulley, and a wheel and axle make work easier?

4. How do some animals use parts of their body as a simple machine?

5. **Cause and Effect**
   You lift your milk carton to take a sip of milk at lunch. Explain how your action caused you to do work.

Lifting Objects With a Lever

Process Skills

Process Skills

- estimating and measuring
- observing
- inferring

Materials

- masking tape
- 3 pencils
- 2 plastic cups
- half-meter stick
- marble
- gram cubes

Getting Ready

In this activity you will find out how moving a fulcrum farther from an object changes the amount of effort needed to lift the object.

Follow This Procedure

1 Make a chart like the one shown. Use your chart to record your observations.

Position of fulcrum	Number of gram cubes	Drawings of observations
25 cm		

2 Tape the three pencils together as shown (Photo A). This will be the fulcrum.

3 Use tape to attach a plastic cup to each end of the half-meter stick. Place the half-meter stick on the pencils. Move the half-meter stick so that the 25 cm mark is directly over the fulcrum (Photo B). The half-meter stick will act as a lever.

4 Put a marble into the cup at the 0 end of the lever. This end of the lever should now be touching the table.

Photo A

Photo B

5 Now **measure** the effort required to lift the cup with the marble. Add gram cubes, one at a time, to the other cup until the cup with the marble is lifted. Record the number of gram cubes you used to lift the marble. Draw what you **observe.**

6 Remove the gram cubes from the cup. Move the fulcrum 5 cm further away from the cup with the marble (Photo C). Record the position of the fulcrum under the half-meter stick in centimeters. Repeat step 5.

Self-Monitoring
Did I put the gram cubes in one at a time until the marble was lifted? Do I have questions to ask before I continue?

Interpret Your Results

1. When you moved the fulcrum farther from the marble, did you use more gram cubes or fewer gram cubes to lift the marble?

Photo C

2. Make an **inference.** If you moved the fulcrum even further from the marble would you need more or fewer gram cubes to lift it?

Inquire Further

How can you test your inference about moving the fulcrum and lifting the marble with gram cubes? Develop a plan to answer this or other questions you may have.

Self-Assessment

- I followed instructions to use a lever to lift a marble.
- I **measured** the number of gram cubes needed to lift the marble.
- I recorded my measurements.
- I made drawings of my **observations.**
- I made an **inference** about placement of the fulcrum and amount of effort needed to lift a marble.

Chapter 2 Review

Chapter Main Ideas

Lesson 1

• A force is a push or a pull that can change the way an object moves.
• Friction is a force that makes an object slow down or stop.

Lesson 2

• Gravity is a force that pulls objects toward the earth.
• Magnetism is the pushing and pulling force of magnets.

Lesson 3

• Work is done whenever a force makes an object move.
• Levers, inclined planes, and screws are simple machines that help you do work and make work easier.
• Wedges, wheel and axles, and pulleys help make the work you do easier.
• Some animals use parts of their body the same way people use simple machines.

Reviewing Science Words and Concepts

Write the letter of the word or phrase that best completes each sentence.

a. force
b. friction
c. fulcrum
d. gear
e. gravity
f. inclined plane
g. lever
h. magnetism
i. pole
j. pulley
k. screw
l. simple machine
m. wedge
n. wheel and axle
o. work

1. A ramp and a slide are examples of an ___.
2. The force of ___ pulls objects toward the center of the earth.
3. A simple machine used to push objects apart is a ___.
4. Magnets produce a force called ___.
5. A ___ is a simple machine that moves or turns an object.
6. You do ___ when you use force to make an object move through a distance.
7. A push or pull is a ___.

8. A ___ is a wheel with jagged edges.

9. Six different ___ can help make work easier.

10. A lever moves back and forth on a ___.

11. A bar magnet has a north and a south ___.

12. A ___ is a simple machine used to hold objects together.

13. The force that slows down or stops moving objects is called ___.

14. A ___ is a simple machine made of a rope or chain and a wheel.

15. A ___ can be used with a fulcrum to help you move an object.

Explaining Science

Draw and label a diagram or write a sentence to answer these questions.

1. How can using more force change an object's speed?

2. How does gravity affect an object?

3. Give two examples of when work is done. For each example, explain why work has been done.

Using Skills

1. You use force to throw a bowling ball down the bowling lane. The ball moving on the lane **causes** friction. What might be the **effect** of the friction between the bowling ball and the ground?

2. **Infer** why it is easier for you to walk on grass than on ice.

3. You see a bird flying by with grass in its beak. Based on your **observation,** explain how the bird is using its beak as a simple machine.

Critical Thinking

1. You are swinging higher than your friend on the swing set. **Draw a conclusion** about who is doing more work. Explain your reasoning.

2. A baseball player hits a baseball high into the sky. **Infer** what will happen to the ball next.

3. Suppose you are trying to move an object using a lever. **Predict** where the fulcrum can be moved to lift the load most easily.

4. **Compare and contrast** an inclined plane and a screw.

Energy To Go!

Cars need energy to move. Energy to move these models comes from batteries. You need energy too. You use it to move, grow, and push the buttons on the remote control. Where does your energy come from?

Chapter 3
Energy in Your World

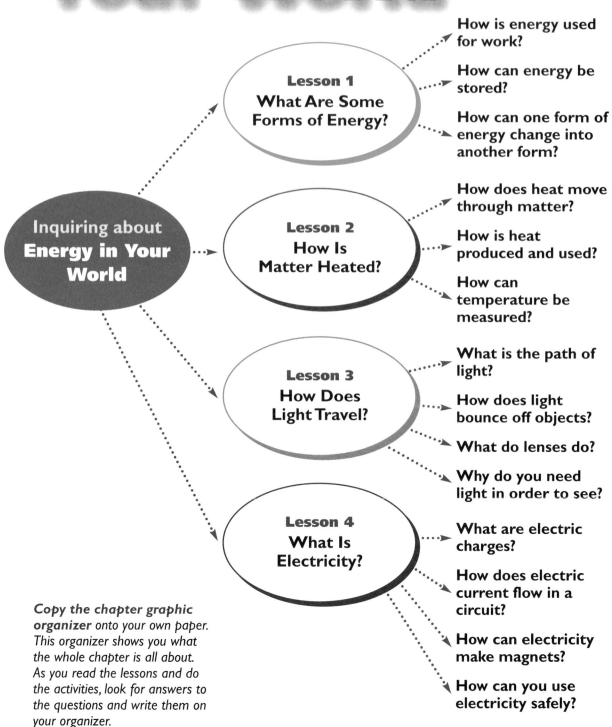

Inquiring about **Energy in Your World**

Lesson 1
What Are Some Forms of Energy?

How is energy used for work?

How can energy be stored?

How can one form of energy change into another form?

Lesson 2
How Is Matter Heated?

How does heat move through matter?

How is heat produced and used?

How can temperature be measured?

Lesson 3
How Does Light Travel?

What is the path of light?

How does light bounce off objects?

What do lenses do?

Why do you need light in order to see?

Lesson 4
What Is Electricity?

What are electric charges?

How does electric current flow in a circuit?

How can electricity make magnets?

How can you use electricity safely?

Copy the chapter graphic organizer onto your own paper. This organizer shows you what the whole chapter is all about. As you read the lessons and do the activities, look for answers to the questions and write them on your organizer.

Exploring Forms of Energy

Process Skills

- observing
- communicating

Materials

- flashlight
- 2 D-cell batteries

Explore

1 The batteries contain materials that store energy. Remove the top of the flashlight. Place the batteries in the flashlight. Replace the top.

2 Hold your hand near the flashlight bulb without turning on the flashlight. **Observe** how the flashlight feels.

3 Turn on the flashlight. Electricity is now flowing through the bulb. What change can you see? Record your observations.

4 Place your hand near the lit bulb. What change can you feel? Record you observations. Then turn off the flashlight.

Reflect

1. In this activity you observed forms of energy such as electricity, light, and heat. Describe the energy changes you observed when you turned on the flashlight.

2. Communicate. Discuss your ideas with the class.

? Inquire Further

What examples of electricity, heat, and light can you find in your classroom? Develop a plan to answer this or other questions you may have.

Measuring Temperature

A thermometer measures temperature. The units of temperature are degrees (°) of Celsius (C) and Fahrenheit (F). You will learn how to estimate and compare temperature in degrees Celsius (sel′sē əs) and Fahrenheit (far′ən hit).

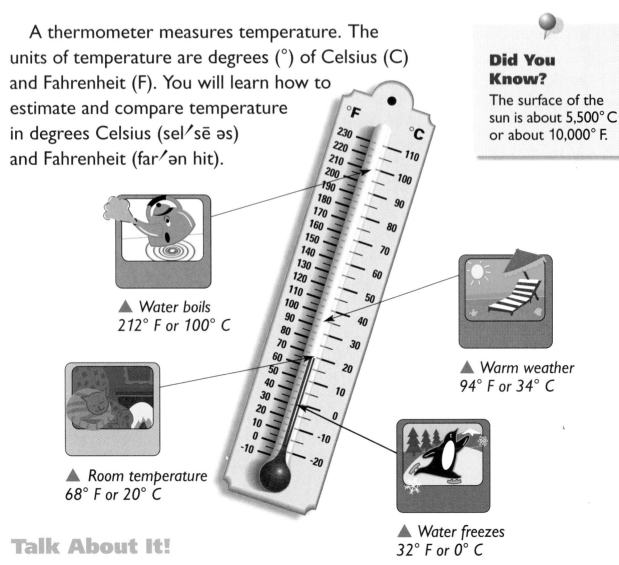

▲ Water boils
212° F or 100° C

▲ Room temperature
68° F or 20° C

▲ Warm weather
94° F or 34° C

▲ Water freezes
32° F or 0° C

Talk About It!

Is 25° C warmer or cooler than 14° C?

Check

Write each temperature using °C.

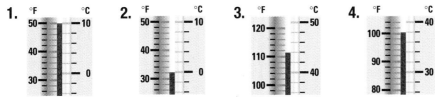

1. °F °C
   50 — 10
   40 —
   30 — 0

2. °F °C
   50 — 10
   40 —
   30 — 0

3. °F °C
   — 50
   120 —
   110 —
   100 — 40

4. °F °C
   — 40
   100 —
   90 —
   80 — 30

You will learn:

- how energy is used for work.
- how energy can be stored.
- about ways that energy changes from one form to another form.

Glossary

energy (en′ər jē), the ability to do work

Lesson 1

What Are Some Forms of Energy?

Beep! The alarm on your clock radio goes off and you jump out of bed. Maybe today you pull the covers over your head instead. You try to go back to sleep. Either way, you're using energy.

Energy and Work

Life Science

Your body uses energy all the time, even when you're asleep. When you breathe, turn over in bed, or blink your eyes, you are using energy.

You recall that work is done when a force makes an object move. Anything that can do work has **energy**. Work includes pulling up the bed covers as well as pushing a swing. Any time you do work, you use energy.

Sometimes you use energy even if you are not doing work. If you push on a heavy box that doesn't move, you are not doing work. You are using energy, though, when you try to make the box move.

Energy is all around you. There are many forms of energy, such as light, sound, and electricity. The brightly lit buildings in the picture use electrical energy. The moving cars have energy that comes from fuel.

Where does electrical energy come from? Where does the energy in fuel come from? Most of the energy we use comes from the sun. Without sunlight, there would be no plants. Without plants, there would be no animals, and most of the fuels we use come from plants and animals. The sun's energy is really the starting point for most of the forms of energy that we use.

Energy from electricity changes to light, which makes the street look bright, even at night. What other things in this scene have energy? ▼

Stored Energy

The man holding the child's swing has just done work. He pulled back on the swing. The swing has stored energy. It is not moving. When the man lets the swing go, the stored energy will change to energy of motion. Energy that can change later to a form that can do work is **stored energy.**

Chemical energy is a form of stored energy. Food, fuel, and batteries store energy. They go through chemical changes. During these changes, energy is given off.

The chemicals that make up food store the energy you need. Chemical changes take place in your body. The energy stored in food becomes energy you need to work and grow. A flashlight battery also stores energy. When you turn on the flashlight, chemical changes in the battery produce electrical energy to light the bulb. Fuel for a car also has stored chemical energy. When the fuel burns, energy is given off. The energy makes the car work.

Stored energy can change into energy of motion. ▼

In your body, food goes through chemical changes that give off energy. You need this energy to keep you warm and move your muscles. ▶

Changing Forms of Energy

You have learned that while the swing is held up high, it has stored energy. When the swing moves down, the stored energy changes to energy of motion. All moving things have **energy of motion.** You store energy when you turn the key in a wind-up toy. The energy changes to energy of motion when you let the toy go and it begins to move.

Other forms of energy can change too. Energy of motion can produce light, as in the bicycle light. Energy from electricity can be changed into different forms of energy. When you turn on a TV, electricity changes to light energy, sound energy, and heat. In fact, all energy you use changes into some other form of energy. Think of some machines that change electricity to other forms of energy.

Glossary

energy of motion
(mō′shən), energy that moving objects have

Glossary

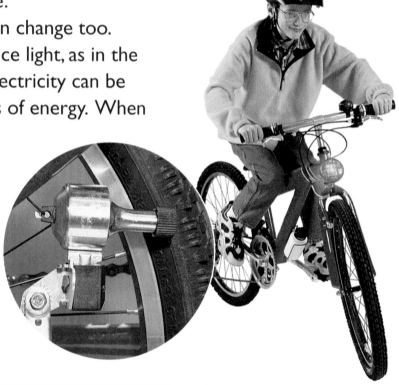

▲ *Energy of motion can change to light. The moving bicycle wheels have energy of motion. As they turn, they produce energy for the generator. The generator provides energy for the light.*

Lesson 1 Review

1. How do we use energy?

2. What is stored energy?

3. What are some examples of one form of energy changing into another form?

4. **Cause and Effect**
   What stores the energy to light a flashlight bulb when the flashlight is turned on?

You will learn:

• how heat moves through matter.

• how heat is produced and used.

• how temperature is measured.

Lesson 2

How Is Matter Heated?

Brrr! It's cold out today. What a day to forget your gloves! Your hands get really cold on your way home. Imagine that a cup of hot soup is waiting for you. How can heat from that cup warm your hands?

How Heat Moves Through Matter

When you hold a warm cup in your cold hands, you feel heat energy moving. Energy is moving from the cup to your hands. Heat energy moves from a warmer object to a cooler one.

Heat comes from many places. When you rub your hands together, you feel heat from friction. You've felt heat coming from a stove and from a fire. What are some other sources of heat?

The boy holding the ice cube is a source of heat. Energy from his body is flowing to the ice cube. His hand feels cold because it is losing heat.

When you hold an ice cube in your hand, energy moves from your hand to the ice. Adding heat energy to the ice makes it melt. ▼

Heat can move through some materials more easily than through other materials. A good **conductor** is a material that allows heat energy to move easily through it. The metal pan in the picture is a good conductor of heat. Have you ever left a metal spoon in a bowl of hot soup? If you have, you know that the metal spoon becomes warm. The spoon is a good conductor of heat.

Some materials are not good conductors. The wooden handle on the pan and the cloth potholder are not good conductors. The wood will not become very warm. Liquids and gases are not good conductors either. Materials that are not good conductors are insulators. An **insulator** is a material that does not allow energy to move easily through it.

Because energy does not move easily through insulators, they help keep hot things hot and cold things cold. Have you ever used a plastic picnic cooler? Did you ever carry a drink in a thermos? These are insulators that some people use every day. Your home probably has a type of insulator in the walls and ceiling. These insulators help you stay comfortable indoors.

Glossary

conductor
(kən duk′tər), a material through which energy flows easily

insulator
(in′sə lā′ tər), a material through which energy cannot flow easily

Trace the flow of heat. It moves from the stove to the pan. From there, the heat flows to the sandwich and melts the cheese. ▼

Glossary

fuel (fyü′əl), a material that is burned to produce useful heat

Making and Using Heat

Heat can be produced when a chemical change takes place in matter. Heat also can be produced by other forms of energy. The pictures show some ways that heat is produced and some things that heat energy can do.

▲ Matter and Heat

*When chemical changes happen, heat is often given off. When fuels such as wood, coal, oil, and gas are burned, heat is produced. A material that is burned to produce heat is a **fuel**.*

◀ Using Heat to Change Matter

Adding heat can make physical and chemical changes happen faster. You might have noticed that sugar dissolves faster in hot tea than in iced tea.

▲ Heat and Light

The light in an electric bulb results from heat. Electric energy makes the wire so hot that it glows, producing light.

◀ Electricity to Heat

Sometimes fuels are used to produce electric energy. Electric energy and some other forms of energy also can give off heat. Tools such as hair dryers change electricity into heat.

Measuring Temperature

What is the temperature where you are today? **Temperature** is a measure of how hot a place or object is. A thermometer is a tool that measures temperature. Hot things or places have high temperatures. Cool things have low temperatures.

Most things get larger, or expand, when they get warmer. Many thermometers work because a liquid inside them expands as the temperature goes up. A temperature scale is marked on the outside of the thermometer. The lines on the scale stand for units known as degrees. The symbol for degrees is °.

Notice the Celsius temperature scale on the thermometer in the picture. Water boils at 100° Celsius, usually written as 100°C. At 0°C, water freezes. The temperature of your classroom is probably about 20°C. Someone can control the temperature inside your school by using a tool called a thermostat.

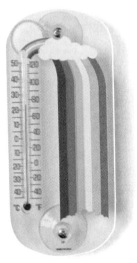

▲ *What is the temperature of the room in which this thermometer is hanging?*

▲ **Thermostat**
This tool controls the temperature inside a building. If the temperature falls below a certain degree, the thermostat turns on a heater.

Lesson 2 Review

1. How does heat travel through matter?

2. What are some ways to produce heat?

3. How is temperature measured?

4. Temperature
Jim and Jane poured cold juice from the same carton into bottles. At lunch time, the temperature of Jim's juice is 7°C. Jane's juice is 12°C. Which bottle is a better insulator?

Making a Thermometer

Process Skills

- making and using models
- observing

Materials

- clear plastic bottle
- funnel
- water (room temperature)
- metric ruler
- red food coloring
- plastic straw
- clay
- container of warm water
- container of cold water

Getting Ready

In this activity you can make your own thermometer and see how it works.

Follow This Procedure

① Make a chart like the one shown. Use your chart to record your observations.

	Observations
Bottle placed in warm water	
Bottle placed in cold water	

② Use the funnel to pour room temperature water into the bottle until the water level is about 4 cm from the bottom of the bottle. Remove the funnel. Add two drops of red food coloring (Photo A).

Photo A

③ Place the straw in the bottle. The straw should be in the water but not touching the bottom of the bottle.

④ Put the clay around the straw to seal the top of the bottle (Photo B). You have made a **model** thermometer.

Photo B

⑤ Hold the base of the bottle in the container of warm water. Do not squeeze the bottle. **Observe** the water in the straw until you see a change (Photo C). Record your observations.

⚠️ **Safety Note** *Be sure to wipe up any spills immediately.*

⑥ Repeat step 5, but hold the bottle in the container of cold water.

Photo C

Interpret Your Results

1. Remember that heat can make things expand. What happened to the water in the thermometer when you placed the thermometer in warm water? Explain.

2. What happened to the water in the thermometer when you placed the thermometer in the cold water? Explain.

⟡ Inquire Further

What would happen if you heated the air in the thermometer model by holding your hands on the sides of the model? Develop a plan to answer this or other questions you may have.

Self-Assessment

- I followed instructions to make a **model** of a thermometer.
- I **observed** the thermometer placed in warm and cold water.
- I recorded my observations.
- I explained what happened to the thermometer when it was placed in warm water.
- I explained what happened to the thermometer when it was placed in cold water.

You will learn:

- about the straight path of light.
- how light bounces off objects.
- what lenses do.
- why sight depends on light.

Glossary

ray, a thin line of light

Lesson 3

How Does Light Travel?

Click! You've just taken a picture of your friends. If you had enough light, you'll have a terrific photograph. You just hope that nobody was making a goofy face.

The Straight Path of Light

Light can come from many sources. Light usually travels in all directions from its source. Think about a light bulb that is lit. Light rays come out all around the bulb. A **ray** is a thin line of light. Rays of light travel in straight lines.

Light can pass through some materials, such as glass and water. It cannot pass through other materials, such as wood, metal, and even you. What happens when light shines on an object that it cannot pass through? You see what happens all the time. A shadow like the one to the left forms.

◀ The shadow forms because rays of light travel in straight lines and cannot bend around the hand. Try this yourself and make the shadow larger and smaller.

Light Bounces

Suppose you tried to read this book in a dark room. You wouldn't be able to see it. We can see things only when light strikes them.

The book is not a light source. When you turn on a lamp, you can see the book. Light from the lamp travels in a straight line to the book. Then the light bounces off the page and into your eyes. Light **reflects** when it bounces off an object. If an object is not a light source, we can see it only when it reflects light to our eyes.

Some objects reflect light very well. Think about a mirror or a piece of silverware. These objects are alike in one way. They have smooth, shiny surfaces. You see yourself in a mirror because it reflects light that strikes your face. The mirror below is reflecting a light beam.

Glossary

Glossary

reflect (ri flekt′), to bounce back

◀ **Reflected Light**
Notice how the light beam travels. As the light hits the mirror, it is reflected onto the objects on the table.

Glossary

lens (lenz), a piece of material that bends light rays that pass through it

▲ *Tiny clues cannot escape this student's search.*

What Lenses Do

If you have a flat piece of glass, light passes straight through it. A window has flat glass. A **lens** is a curved piece of clear material. Light rays bend when they pass through a lens.

Lenses make things look different from how they usually appear. Lenses can make things appear bigger or smaller. Lenses have different shapes, and we use them for different purposes. The picture below shows lenses and some of the tools that use them.

Some lenses are thick in the middle and thin at the edges. These lenses are often used to make things appear larger. Other lenses are thin in the middle and thick at the edges. These lenses are often used to make things appear smaller. Both kinds of lenses are used in eyeglasses to help people see better. What shape is the lens that the student is using?

◀ *Lenses have many uses. In a camera, the lens makes a small image, or picture, from a large object. Field glasses have several lenses to make things appear larger. What other things use lenses?*

Light and Sight

Human Body

Parts of your eyes bend light rays so you can see. The picture shows some of these parts. The outer part of each eye has a clear covering. This is the first part of the eye to bend light. Trace the path that light takes through the eye.

Behind the clear covering is a small opening called the pupil. You can see this opening as the black circle in the center of your eye. Around the pupil is the iris, which is the colored part of your eye.

Behind the pupil and the iris is the lens. It bends light rays much as a glass or plastic lens does. After light passes through the lens, an image of what you see forms on the back of your eye. From there, a message travels to your brain. Your brain tells you what you see.

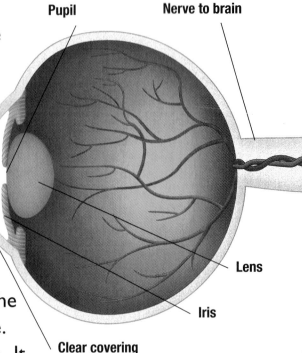

Pupil

Nerve to brain

Lens

Iris

Clear covering

Lesson 3 Review

1. How does light travel from a source?

2. How does light bounce off an object?

3. What do lenses do?

4. Why does sight depend on light?

5. Sequencing
Put these events in the proper order:
a. light enters the lens of the eye; **b.** light is reflected from an object; **c.** light ray leaves a light source; **d.** an image forms on the back of the eye.

You will learn:
- about electric charges.
- how electric current flows in a circuit.
- about electricity and magnets.
- how to use electricity safely.

Glossary

electric charges (i lek′trik chär′jəz), tiny amounts of electricity present in all matter

Lesson 4

What Is Electricity?

Power's out! The storm broke the electric lines. You have no electricity. It's totally dark. You might see sparks in your clothes as you move around. Those sparks are electricity. Where did they come from?

Electric Charges

Electricity is a form of energy. You use it all the time. It lights your home. It lets your radio and TV bring you sound and pictures. It runs many machines. Electricity is all around us. Electricity is even in our bodies.

You might sometimes see sparks when you take your clothes off in a dark room. All matter has very tiny amounts of electricity known as **electric charges**. The sparks are electric charges moving from one object to another. Suppose you rub your feet on a rug and then touch a doorknob or pet a cat, as this boy is doing. You might get a shock. Electric charges are moving again!

◀ You might get a shock when you pet a nice, furry animal. Electric charges are jumping between you and the animal.

Your body, like all matter, has electric charges. When you rub your feet on something such as a rug, you pick up more electric charges. Those extra charges might then move to something else. When the charges move, you might get a shock.

The girl in the picture is rubbing a balloon against a wool sweater. Electric charges are moving from the sweater to the balloon. The balloon becomes charged, which means it picks up electric charges from the wool.

The charged balloon can stick to the wall. The balloon sticks because it and that part of the wall have unlike, or opposite, electric charges. Things with unlike charges attract, or pull on, each other. The pulling force between the unlike charges makes the balloon stick to the wall.

Try rubbing two balloons on the same piece of wool. Then both balloons would have the same kind of charge. They would push away from each other. Things that have the same kind of charge push apart.

Getting a Charge
Rubbing a balloon on wool gives it an electric charge. The part of the wall near the balloon has an opposite charge, so the balloon sticks to it. ▼

Glossary

electric current
(kėr′ənt), the flow of electric charges

electric circuit
(sėr′kit), the path along which electric current moves

Electric Current and Circuits

Even though electricity is all around us, we need to control it in order to make it do useful work. Electricity that we use moves inside wires and flows from place to place. The flowing movement of electricity from place to place is an **electric current.**

Electric power plants make electric current. That electric current, or electricity, travels along wires into our homes. A battery can also provide electric current. Follow the path of the current in the picture. Notice that the path goes all around. The path through which electric current flows is an **electric circuit.**

A Circuit in a Flashlight

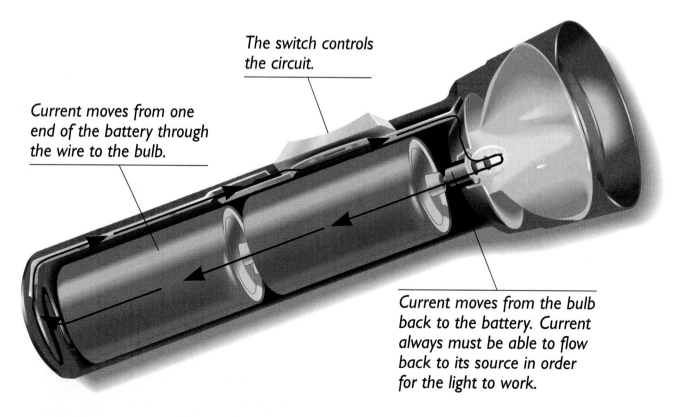

The switch controls the circuit.

Current moves from one end of the battery through the wire to the bulb.

Current moves from the bulb back to the battery. Current always must be able to flow back to its source in order for the light to work.

Electric current can flow through wires only when a circuit is complete. If there is a break in the circuit, electricity cannot flow. The switch in a flashlight opens and closes the circuit. The switches for lights in a room work in much the same way.

A Circuit in a Room

◀ The light switch in a room controls the flow of current in a circuit. When the switch is down, in the "off" position, there is a break in the circuit. Current cannot flow. The light is off.

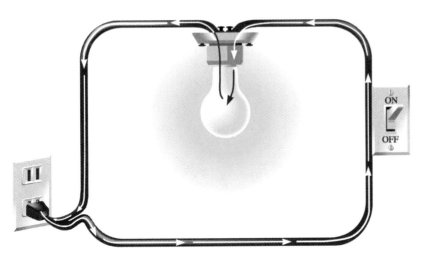

◀ Turn the switch up, to the "on" position. Now there is no break in the circuit. Electric current can flow to the bulb. The light is on.

Glossary

electromagnet
(i lek′trō mag′nit), a metal that becomes a magnet when electricity passes through wire wrapped around it

Magnets and Electricity

Electric current can produce magnetism. Look at the magnet that the student made. She wrapped plastic-coated wire around an iron bolt. The ends of the wire are attached to a battery. When electric current flows through the wire, the bolt becomes a magnet. This magnet is an **electromagnet.**

You may be surprised to learn that you use electromagnets every day. They are used in doorbells, telephones, microphones, and electric motors. The picture below shows a big, powerful electromagnet. It picks out metal objects from piles of trash. The metal is taken away later and made into new objects.

If the student takes a wire off the battery, she breaks the circuit. The bolt stops being an electromagnet after a while. ▼

A crane lifts the magnet and the metal that is stuck to it. When the magnet is above the place where metal is piled up, the electricity is turned off. Then the pieces of metal drop off the magnet and fall onto the pile. ▼

Using Electricity Safely

Electricity can be harmful. You must always use it carefully. The chart shows some ways you can be safe while using electricity.

Using Electricity Safely

- Make sure hands are dry when you touch electric switches and appliances.

- Never stand on a wet floor while using electricity.

- Keep electric appliances away from water.

- Keep electric cords in places where they will not be damaged.

- Do not use appliances that have worn out or broken cords or plugs.

Lesson 4 Review

1. How can matter get an extra electric charge?

2. How does a circuit carry electric current?

3. How can electricity make a magnet?

4. What are some ways to use electricity safely?

5. **Cause and Effect**
   How can you cause an iron bolt to become an electromagnet?

Making an Electric Circuit

Process Skills

- observing
- inferring
- making operational definitions

Materials

- safety goggles
- piece of aluminum foil
- scissors
- masking tape
- D-cell battery
- flashlight bulb
- clothespin

Getting Ready

You will be building an electric circuit. Aluminum foil will take the place of the wires that are in many circuits.

Follow instructions carefully to see if you can get the bulb in the circuit to light up.

Follow This Procedure

1 Make a chart like the one shown. Use your chart to record your observations.

	Observations of bulb
Base of bulb touching foil	
One foil strip removed from battery	

2 Put on your safety goggles. Fold the aluminum foil the long way. Then fold it again four more times. Cut the strip in half (Photo A).

Safety Note *Be careful when folding and handling foil strips. There may be sharp edges.*

3 Tape one end of one strip to one end of the battery. Tape one end of the other strip to the other end of the battery (Photo B).

4 Wrap the loose end of one foil strip around the base of the flashlight bulb. Hold the strip in place with the clothespin (Photo C).

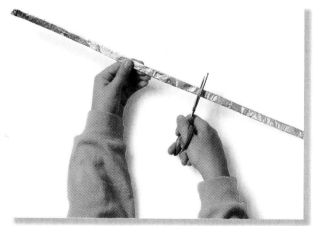

Photo A

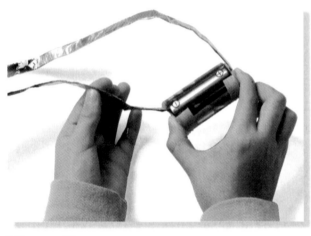

Photo B

 Safety Note If the battery and foil become warm, disconnect the circuit and allow them to cool.

⑤ Touch the base of the bulb to the end of the other foil strip. Be sure no parts of the two strips touch each other. Look at the bulb. Record your **observations.**

Self-Monitoring
Did the bulb light? If not, did I make sure that no parts of the foil strips were touching each other?

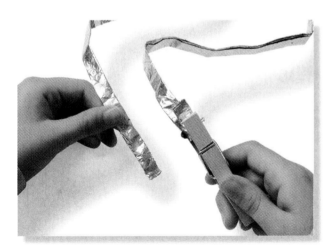

Photo C

⑥ Remove the foil strip attached to the light bulb from the battery. Touch the base of the bulb to the end of the other foil strip. Look at the bulb. Record your observations.

Interpret Your Results

1. Did the bulb light up when both strips were attached to the battery? Did the bulb light up when you removed one of the strips from the battery? What can you **infer** from your observations about an electric circuit?

2. Write an **operational definition** of an electric circuit.

Inquire Further

How could you find out all the ways you use circuits in one day? Develop a plan to answer this or other questions you may have.

Self-Assessment

- I followed instructions to make an electric circuit.
- I **observed** the light bulb.
- I recorded my observations.
- I made **inferences** based on my observations of the electric circuit.
- I wrote an **operational definition** of an electric circuit.

Chapter 3 Review

Chapter Main Ideas

Lesson 1

• Energy exists in many forms and is used whenever work is done.

• Energy can be stored and can change to a form that can do work.

• Forms of energy can change.

Lesson 2

• Heat moves from a warm place or object to a cooler place or object.

• Forms of energy can produce heat that we use, and heat can change into other forms of energy.

• A thermometer is a tool for measuring temperature, which tells how hot an object or place is.

Lesson 3

• Light travels in straight lines from its source.

• Light can bounce off objects.

• A lens can bend light rays.

• You see objects when light enters your eye.

Lesson 4

• All matter has electric charges.

• Electricity flows in a circuit.

• Electricity flowing in a coiled wire wrapped around a piece of metal can make a magnet.

• You can use electricity safely if you use it correctly.

Reviewing Science Words and Concepts

Write the letter of the word or phrase that best completes each sentence.

a. conductor
b. electric charges
c. electric circuit
d. electric current
e. electromagnet
f. energy
g. energy of motion
h. fuel
i. insulator
j. lens
k. ray
l. reflect
m. stored energy
n. temperature

1. The ability to do work is ___.
2. Energy that can change into a form that can do work is ___.
3. The energy that moving objects have is ___.
4. A material through which energy flows easily is a ___.
5. A material through which energy cannot flow easily is an ___.
6. A material that is burned to produce useful heat is a ___.

7. The ___ of a place or object is a measure of how hot it is.

8. A thin line of light is a ___.

9. To ___ is to bounce off an object.

10. A piece of material that bends light rays that pass through it is a ___.

11. Tiny amounts of electricity present in all matter are ___.

12. The flow of electric charges is an ___.

13. The path along which electric current moves is an ___.

14. A metal that becomes a magnet when electricity passes through wire wrapped around it is an ___.

Explaining Science

Draw and label a diagram or write a paragraph to answer these questions.

1. What is one way that stored energy can change to become energy of motion?

2. How are conductors different from insulators?

3. How is the path of light hitting a mirror different from the path of light through a lens?

4. How does electric current move through a turned-on flashlight?

Using Skills

1. Suppose you **measure** the **temperature** outside as 34°C. Would you need to wear a coat?

2. A plastic or rubber coating is around electric wires. What do you **infer** is the purpose for this coating?

3. **Make a generalization** about why electromagnets are useful.

4. What do you **conclude** happens to electric energy when you turn on an electric lamp?

Critical Thinking

1. **Apply** what you learned to describe what happens to electric energy when you turn on a radio.

2. What can you **conclude** happens when your hand gets cold from touching a cold object?

It's a One-Person Band!

This one instrument can make many different sounds. The sounds played together make music. You constantly hear sounds. How is sound important to you? You might be surprised.

Chapter 4
Sound

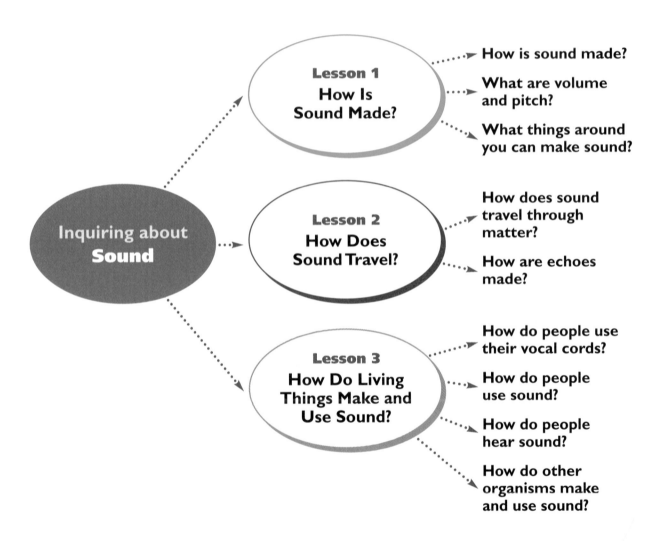

Inquiring about Sound

Lesson 1
How Is Sound Made?
- How is sound made?
- What are volume and pitch?
- What things around you can make sound?

Lesson 2
How Does Sound Travel?
- How does sound travel through matter?
- How are echoes made?

Lesson 3
How Do Living Things Make and Use Sound?
- How do people use their vocal cords?
- How do people use sound?
- How do people hear sound?
- How do other organisms make and use sound?

Copy the chapter graphic organizer onto your own paper. This organizer shows what the whole chapter is all about. As you read the lessons and do the activities, look for answers to the questions and write them on your organizer.

Exploring Sound

Process Skills

- observing
- communicating

Materials

- safety goggles
- plastic wrap
- soup can
- rubber band
- salt
- baking tray
- half-meter stick
- cardboard tube

Explore

1 Put on your safety goggles. Stretch the plastic wrap over the open end of the can. Use the rubber band to hold the plastic in place.

2 Sprinkle some salt on top of the plastic.

3 Hold a baking tray near the top of the can and tap it with the half-meter stick. Tap the tray softly, then harder. Record your **observations.**

4 Try to make the salt jump with your voice. Point the cardboard tube so that it is just above the can but not pointed directly at the salt. Call through the tube softly, then louder. Do not blow on the salt. Record your observations.

Reflect

Communicate. Describe how the sounds you made affected the salt.

Inquire Further

What other sounds can make the salt move? Develop a plan to answer this or other questions you may have.

Drawing Conclusions

Think of the many clues you use to decide what is going on around you. Maybe you hear a voice or a sound that you have heard before. Or maybe you see something that you have already seen. You might use clues like these to make a decision, or **draw a conclusion**. A conclusion is a decision that you make based on evidence and reasoning. As you read Lesson 1, *How Is Sound Made?*, think about the sounds you hear every day. Ask yourself what conclusions you can draw about different sounds.

Reading Vocabulary

conclusion
(kən klü′zhən), a decision or opinion based on evidence and reasoning

Example

You hear a bounce, bounce, thump coming from outside. You have heard those sounds before. You draw the conclusion that someone is playing basketball. Looking through the window, you see that you are correct.

Talk About It!

1. What clues can you use to draw conclusions?

2. Do you think a loud sound or a soft sound is coming from the whistle in the picture? How did you draw the conclusion you did?

▼ *Have you ever wondered why sounds are different from each other?*

B 87

You will learn:

- how sound is made.
- about volume and pitch.
- what things around you can make sound.

Glossary

vibrate (vī′brāt), move quickly back and forth

Lesson 1

How Is Sound Made?

Ring! Bang! Thump! Inside, the doorbell is ringing and a door is being slammed. Outside, a ball is bouncing. It seems like almost everything around you is making some kind of sound.

Making Sound

You know that light and heat are forms of energy. You may have already guessed that sound is also a form of energy.

The girl in the picture is making sound by blowing into a whistle. Sound is made when matter vibrates. To **vibrate** means to move quickly back and forth. You can feel vibrations if you touch a bell that is ringing or a radio that is playing. The sounds you hear may be different, but they are all alike in one way. All sounds are made by vibrating matter.

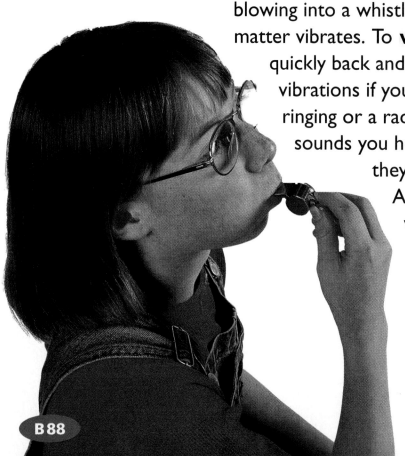

◄ The girl is blowing air into a whistle. The air vibrates as it passes through the opening at the top of the whistle. You hear the sound of the whistle.

Volume and Pitch

Stop and think about the different sounds you hear every day. Some are soft, some are loud. Sounds can have different volumes. **Volume** is the loudness or softness of a sound. Loud sounds have more volume than soft sounds. Imagine whispering a secret to a friend. Your whisper has a lower volume than your normal talking voice. You change the volume of your voice all the time.

Two sounds can have the same volume but a different pitch. **Pitch** is how high or low a sound is. Objects that vibrate slowly have a low pitch. Objects that vibrate quickly have a high pitch. Suppose you hit each of the instruments on this page. The triangle would have a high pitch because it vibrates quickly. The gong would have a low pitch because it vibrates slowly.

A gong has a low pitch because it vibrates slowly. ▶

Glossary

volume (vol′yəm), the loudness or softness of a sound

pitch (pich), how high or low a sound is

Glossary

▲ A triangle has a high pitch because it vibrates quickly.

Things Around You That Make Sound

Imagine a day with no sounds happening anywhere. No sounds would mean no matter is vibrating anywhere around you. That is impossible! There are phones ringing, people talking, and shoes squeaking. All these sounds have different volumes and different pitches. Listen right now. You just cannot get away from sounds. Read about different sounds on these two pages.

▲ **Helpful Sounds**
Think of what the sound of the siren coming from this fire truck might mean. Now think of other sirens you have heard and what the sound of each siren might have meant.

Sounds at the Park
There can be many different kinds of outside sounds. Imagine you were visiting this park. You would hear different voices. You might hear the equipment squeak or the sound of feet jumping on the ground. Think of other sounds you might hear at the park. ▼

◀ Sounds in the Forest

It looks like it would be very quiet if you took a walk through the forest in this picture. But if you listened carefully, you would probably hear many different sounds. Maybe you would hear the leaves rustling in the breeze or the owls hooting. What other sounds might you hear in the forest?

▲ Sounds at Home

Sounds can come from different sources in this room. Some sounds might have a high volume. Other sounds might have a low volume. Think of how you could change the volume of the sounds.

Lesson 1 Review

1. How is sound made?

2. What are two ways sounds can be different?

3. Name two things around you and describe how they make sound.

4. **Drawing Conclusions**
   You hear the low pitch of a drum and the high pitch of a bell while you are listening to music. What conclusion can you draw about the vibration of each instrument?

Changing Pitch

Process Skills

- observing
- predicting
- inferring

Materials

- safety goggles
- facial tissue box
- 2 rubber bands (different sizes)
- 2 pencils
- half-meter stick

Getting Ready

In this activity you can find out how pitches change when you change the length of vibrating rubber bands.

Follow This Procedure

1 Make a chart like the one shown. Use your chart to record your observations.

	Prediction	Observations
Rubber bands 15 cm apart	X	
Rubber bands 5 cm apart		

2 Put on your safety goggles. Stretch the rubber bands the long way around the tissue box. Place the pencils under the rubber bands 10 cm apart (Photo A).

⚠️ **Safety Note** Handle the rubber bands carefully so they don't snap or break.

3 Pluck the rubber bands between the pencils. Listen to the sounds being produced. Notice how high or low the pitches of the rubber bands sound.

4 Move the pencils so that they are 15 cm apart (Photo B). Pluck the rubber bands. Listen to the pitches of the sounds. How did they change? Record your **observations.**

5 How will the pitches of the sounds change if you move the pencils closer together? Record your **prediction.**

6 Repeat step 4, but move the pencils so that they are 5 cm apart.

Photo A

Photo B

Self-Monitoring
Have I followed the directions correctly and moved the pencils to hear how the pitches change? Do I need to repeat any steps to check my observations?

Interpret Your Results

1. The part of the rubber bands between the pencils vibrated and produced sound. What can you **infer** about the length of the vibrating part of a rubber band and the pitch?

2. Moving the pencils apart made the rubber bands vibrate slower. Moving them close together made the rubber bands vibrate faster. What can you infer about the speed of vibration and the pitch of the sound?

Inquire Further

How do musicians change the pitches of instruments such as violins or guitars? Develop a plan to answer this or other questions you may have.

Self-Assessment

- I followed instructions to change the pitch of sound produced by vibrating rubber bands.
- I made **observations** of the pitches produced.
- I **predicted** how the pitches would change if the pencils were moved closer.
- I recorded my prediction and observations.
- I made **inferences** about the length of a vibrating object, the speed of its vibration, and the pitch of its sound.

You will learn:

- how sound travels through matter.
- how echoes are made.

Lesson 2

How Does Sound Travel?

Rrrrr! You hear a low roaring sound in the sky. That sound is not strange to you. You know it's an airplane. How can you be hearing an airplane that is so far away? How does sound travel that far?

How Sound Travels Through Matter

The instrument is the source of sound. Sound from the instrument travels through the air in all directions. Everyone around the musician can hear it. Sound is loudest close to its source. Sounds get softer as you move farther away from the object making the sound. ▶

You can tell by the look on their faces that the people in the picture can hear the music. How did sound travel to their ears? Remember that objects make sound when they vibrate. Sound travels out in all directions from a vibrating object.

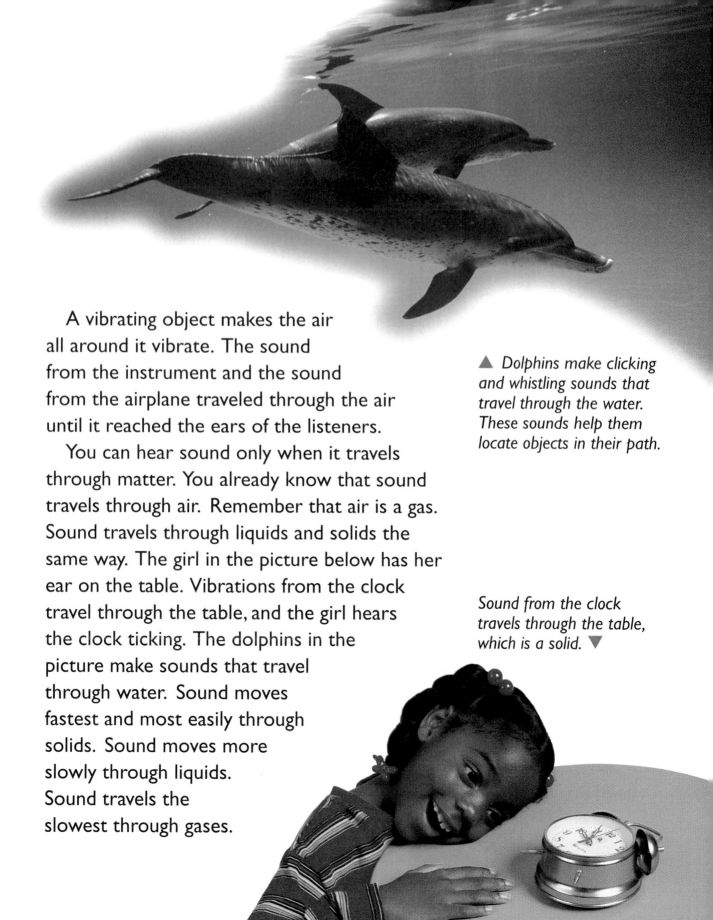

A vibrating object makes the air all around it vibrate. The sound from the instrument and the sound from the airplane traveled through the air until it reached the ears of the listeners.

You can hear sound only when it travels through matter. You already know that sound travels through air. Remember that air is a gas. Sound travels through liquids and solids the same way. The girl in the picture below has her ear on the table. Vibrations from the clock travel through the table, and the girl hears the clock ticking. The dolphins in the picture make sounds that travel through water. Sound moves fastest and most easily through solids. Sound moves more slowly through liquids. Sound travels the slowest through gases.

▲ Dolphins make clicking and whistling sounds that travel through the water. These sounds help them locate objects in their path.

Sound from the clock travels through the table, which is a solid. ▼

Glossary

Glossary

echo (ek′ō), a sound that bounces back from an object

An Echo

Imagine throwing a ball against a wall. The ball hits the wall and bounces back to you. Now imagine shouting in a large, empty room like the one in the picture. Like the ball, your voice bounces off of the walls. You hear an echo. An **echo** is sound bouncing back from an object, as shown in the picture. Your voice traveled through the air and bounced off of the walls of the room. Sound does not always bounce back. Notice the picture of the room with furniture. The furniture takes in, or absorbs, some of the sound, and you do not hear an echo.

▲ You hear an echo when sound bounces off of the smooth, hard walls of an empty room.

▲ You do not hear an echo in a room full of furniture because the furniture absorbs some of the sound.

Sound travels through the air, hits the wall, and bounces back. ▼

Sound can bounce off of different surfaces. You can hear an echo if you are surrounded by hills, cliffs, or large buildings. However, you can hear an echo best when sound bounces off of a hard, smooth surface.

Some animals, like the bat in the picture, use echoes to safely find their way around. Bats hunt for food at night. Most bats cannot see well in the dark. As they fly, they make short, high-pitched sounds. These sounds bounce off of objects that are in the bat's path. They keep the bat from flying into things. Bats also use sounds to find food. The sounds bounce off of insects so the bat knows where they are.

▲ *Echoes help bats find food and keep them from flying into trees and any other objects in the way.*

Lesson 2 Review

1. How does sound travel through different forms of matter?

2. How are echoes made?

3. **Drawing Conclusions**
   You hear your friend calling your name, but you cannot see him. What conclusion can you draw about how the sound of your friend's voice reached your ears?

Listening to Sound Through Matter

Process Skills

- observing
- classifying
- inferring

Materials

- half-meter stick
- metric ruler
- penny
- wood block

Getting Ready

In this activity you will listen to and describe sound traveling through different materials.

Follow This Procedure

1 Make a chart like the one shown. Use your chart to record your observations.

	Volume Soft ⟶ Loud				
Air	1	2	3	4	5
Wood block	1	2	3	4	5

2 Place the half-meter stick on your desk.

3 Place your ear close to the desk near the 30 cm mark on the half-meter stick. Have a partner position the ruler at the 0 point of the half-meter stick. Your partner should use the metric ruler to drop the penny from a height of 10 cm (Photo A). **Observe** the volume of the sound as it travels through the air.

Photo A

Photo B

④ Repeat step 3, but this time press your ear to a block of wood placed at the 30 cm mark of the half-meter stick (Photo B). Observe the volume of the sound as it travels through the wood block.

⑤ **Classify** the sounds on your chart. Your chart has a volume scale. If 1 is the softest volume and 5 is the loudest, how would you rank the volume of sound through air? Circle the number of your choice. Then rank the volume of sound through wood. You may need to repeat the procedure to double-check your observations.

Self-Monitoring
Do I need to repeat any part of the procedure to check my observations?

Interpret Your Results

1. Through which material did the sound seem loudest?

2. Through which material did the sound seem softest?

3. Which do you think would sound louder — sound traveling through a metal block or sound traveling through air? Record your **inference.** Explain your answer.

Inquire Further

How would the volume change if you listened to sound traveling through water? Develop a plan to answer this or other questions you may have.

Self-Assessment

- I followed instructions to listen to sound through different materials.
- I **observed** the volume of sound through air and through the wood block.
- I recorded my observations.
- I **classified** the volume of sound.
- I made an **inference** about how sound travels through different materials.

You will learn:

- how people use their vocal cords.
- how people use sound.
- how people hear sound.
- how other organisms make and use sound.

How Do Living Things Make and Use Sound?

Croak! **Buzz!** Chirp! **Chatter!**
Do animals have conversations?
Well, maybe they don't have
conversations like we do, but many
animals do communicate with each
other using sound.

Glossary

vocal cords
(vō′kəl kôrdz), two small folds of elastic tissue at the top of the windpipe

Using Your Vocal Cords

Place your fingers on your throat like the girl in the picture is doing. Now say your name out loud. There seems to be something vibrating in your throat.

What you feel are your vocal cords vibrating as you speak. **Vocal cords** are thin folds at the top of your windpipe. When you talk, you move air from your lungs past your vocal cords. The air makes your vocal cords vibrate. Your vibrating vocal cords make sound.

◀ *You can feel your vocal cords vibrate when you touch your throat as you talk. This vibration makes sound. You use your mouth, teeth, and tongue to change the sound to form words.*

How You Use Sound

Just like the baby in the picture below, you learned very early to use sound to communicate. Crying, babbling, and cooing were about the only sounds you could make. So you would go through each day crying, babbling, and cooing as a way to tell everyone what you wanted. As you grew older, you learned to control the sounds you made.

Think about the many ways you use sound now in communicating and listening. You use your voice to talk to people directly, as shown in the picture on the right, or on the telephone. Have you ever whistled a song? Perhaps you have listened to a radio for news or music. Think of the last time you heard a siren that warned you of danger.

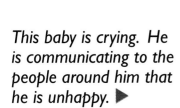

▲ It is good to tell somebody when you need something.

This baby is crying. He is communicating to the people around him that he is unhappy. ▶

Glossary

eardrum
(ir′drum′), the thin, skinlike layer that covers the middle part of the ear and vibrates when sound reaches it

nerve (nėrv), a part of the body that carries messages to the brain

How You Hear Sound

Like the children in the picture, you use your ears to hear what someone is saying to you. You see your ears every time you look in a mirror. You know that they are important in hearing sound. The question is, how do your ears work? Sound travels through the air and goes into them. What happens then? From what you know about sound, you can guess that something is going to vibrate in your ears.

Human Body

The first part of your body that sound reaches is your outer ear. Your ear also has many other parts that are inside your head. Sound travels through both the outside and the inside parts of your ear before messages about the sound reach your brain. Follow the path of sound on the next page as it travels from the air through your ears.

You hear the voice of a person whispering into your ear. ▶

1 Outer Ear

The outer part of your ear is the part you see when you look in a mirror. The outer ear catches sound traveling in the air. It then moves the sound to the part of the ear inside the head.

2 Eardrum

Sound travels from the outer ear to the eardrum. The **eardrum** is the thin, skinlike layer that covers the middle part of the ear. When sound hits the eardrum, the eardrum begins to vibrate.

3 Three Tiny Bones

The vibrating eardrum makes the three tiny bones in the middle of the ear vibrate.

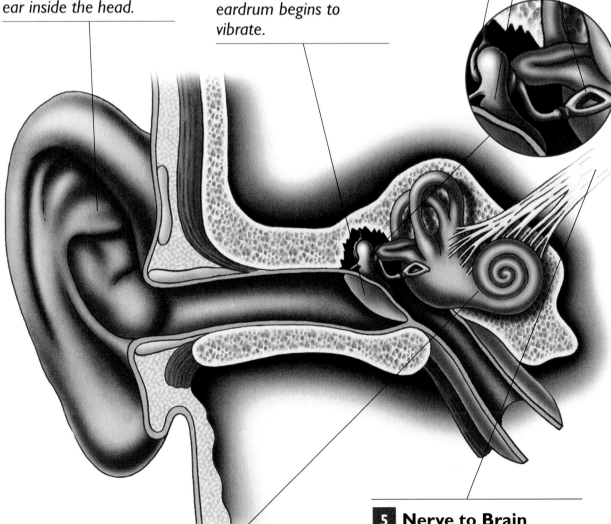

4 Part Shaped Like a Shell

This part of the ear is shaped like the shell of a snail. It is filled with liquid. The vibrating bones cause the liquid to vibrate.

5 Nerve to Brain

The liquid carries sound vibrations to a special nerve. This **nerve** carries messages to the brain. The brain helps you understand the sound you hear.

How Other Organisms Make and Use Sound

Rattlesnakes

A rattlesnake shakes its tail rattle when it thinks it is in danger. The rattle is made of hornlike pieces that are loosely connected. The vibrating tail rattle makes a whirring sound. The rattlesnake is giving a warning. ▼

Life Science

Most sounds that living things make have a meaning. Like people, many other living things use sounds to help them communicate and survive. Some use vocal cords to make sound. A dog uses its vocal cords to bark or growl. Others use different parts of their body. Some rabbits will thump a hind foot on the ground to warn other rabbits of danger. Find out how the living things on these pages make and use sound.

▲ **Grasshoppers**

Grasshoppers do not have vocal cords. Some grasshoppers make sound by rubbing their back legs against their front wings. They use this sound to "sing" to other grasshoppers.

▲ **Frogs**

A frog uses its vocal cords to make sound. The frog forces air from its lungs. The air passes over the frog's vocal cords. The vocal cords vibrate and make sound. Some frogs have a large air sac. The air sac helps the frog make louder sounds than a frog that does not have an air sac. The frog uses sound to call to other frogs or warn of danger.

▲ **Crayfish**

Crayfish, or crawfish, live in and around lakes and rivers. A crayfish has a hard shell and four feelers at the front of its body. The crayfish rubs its feelers against its shell to make sound. It makes this sound to frighten other animals away.

Lesson 3 Review

1. How do vocal cords make sound?

2. Name two ways you use sound.

3. What path do sound vibrations follow through the ear?

4. What are two ways other organisms make and use sound?

5. **Cause and Effect**
   Suppose the eardrum did not vibrate. What effect would this have on a person's hearing?

Chapter 4 Review

Chapter Main Ideas

Lesson 1

• Sound is made when matter vibrates.

• Volume is the loudness or softness of a sound. Pitch is how high or low a sound is. The pitch of a sound depends on how fast an object vibrates.

• Sounds are all around you.

Lesson 2

• Sound travels out in all directions through liquids, solids, and gases.

• Echoes are made when sound bounces back from an object.

Lesson 3

• You make sound when your vocal cords vibrate.

• You use sound to communicate.

• Sound travels from the air through all the parts of your ear. You hear when a special nerve carries messages to the brain.

• Other living things make sound using different body parts, such as wings, legs, and feelers. They use sound to warn of danger, to call to others, or to frighten other animals away.

Reviewing Science Words and Concepts

Write the letter of the word or phrase that best completes each sentence.

a. eardrum
b. echo
c. nerve
d. pitch
e. vibrate
f. vocal cords
g. volume

1. The middle part of the ear is covered by a thin, skinlike layer called the ___ .

2. To make sound, matter must ___ , or move quickly back and forth.

3. Objects that vibrate slowly make sounds with a low ___ .

4. Messages about sound are carried by a special ___ to the brain.

5. A sound that bounces back from an object is called an ___ .

6. Soft sounds do not have as much ___ as loud sounds.

7. When you talk, air moves from your lungs past your ___ .

Explaining Science

Write a sentence to answer these questions.

1. How can the pitch of a sound be changed?

2. Does sound travel fastest through a gas, a liquid, or a solid?

3. How do you make sound using your vocal cords?

Using Skills

1. You hear the telephone start to ring. **Draw a conclusion** about what is happening inside the telephone that causes it to make sound.

2. Does sound travel faster through the outer part of your ear or through the three tiny bones in your ear? **Communicate** your ideas in one or two sentences.

3. Suppose you call out to someone and hear an echo. What can you **infer** about the sound you made?

Critical Thinking

1. Suppose a radio is placed in a room with no air. **Predict** whether sound would travel through the room when the radio is turned on. Explain your prediction.

2. Infer how covering your ears can keep you from hearing sound.

3. Compare and contrast sounds. How are all sounds alike? How can sounds be different?

4. Put the way sound travels from the air through your ear in the correct **sequence:** three tiny bones in the middle of your ear vibrate, your outer ear catches sound and moves it to the part of the ear inside your head, a special nerve carries messages about sound to your brain, your eardrum begins to vibrate, liquid in the part of your ear shaped like a snail's shell starts to vibrate.

Unit B Review

Reviewing Words and Concepts

Choose at least three words from the Chapter 1 list below. Use the words to write a paragraph about how these concepts are related. Do the same for each of the other chapters.

Chapter 1
chemical change
gas
liquid
physical change
solid
states of matter

Chapter 2
force
gravity
inclined plane
lever
magnetism
simple machine

Chapter 3
conductor
electric circuit
electric current
electromagnet
energy
insulator

Chapter 4
eardrum
nerve
pitch
vibrate
vocal cords
volume

Reviewing Main Ideas

Each of the statements below is false. Change the underlined word or words to make each statement true.

1. A <u>mixture</u> is something about an object that you can observe with your senses.

2. Solid, liquid, and gas are three <u>chemical changes</u>.

3. Water <u>condenses</u> when it changes from a liquid to a gas.

4. A push or a pull is a kind of <u>friction</u>.

5. The force of <u>magnetism</u> pulls objects toward the center of the earth.

6. Energy cannot flow easily through a <u>conductor</u>.

7. Tiny amounts of electricity present in all matter are <u>electromagnets</u>.

8. A <u>lens</u> is a path through which electric current moves.

9. Objects that vibrate slowly make sounds with a low <u>volume</u>.

10. Messages about sound travel on a special <u>vocal cord</u> to the brain.

Interpreting Data

The diagram below shows the temperature on thermometers that are different distances from a heat source. Use the diagram to answer the questions below.

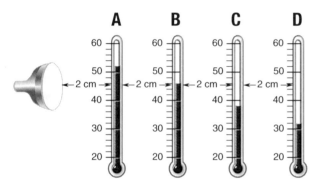

1. The spaces between the lines on the thermometers stand for 2°C. Look at the top of the colored liquid in Thermometer A. What is the temperature? How far is the thermometer from the lamp?

2. What is the temperature on Thermometer B? How far is this thermometer from the lamp?

3. What is the temperature of the thermometer that is 6 cm from the lamp? 8 cm from the lamp?

4. How does temperature change when distance becomes greater between a thermometer and a heat source?

Communicating Science

1. Draw and label a diagram that shows how water changes from one state to another.

2. Draw and label a diagram that explains how a lever can be used to lift a heavy object.

3. Write a paragraph to explain how lenses can make objects look larger or smaller.

4. Write a paragraph to explain how the vocal cords are like the strings on a guitar.

Applying Science

1. Choose five things in your classroom. Make a list of the properties of each of the items, but do not name the items. Exchange your list with a partner and try to guess the items by their properties.

2. Suppose you are making instruments for a class band. You want to make a rubber band "guitar" with a high pitch. Which of the rubber bands in the picture below would make a sound with a higher pitch?

Unit B
Performance Review

Job Fair

Using what you learned in this unit, complete one or more of the following activities to be included in a Job Fair event. These exhibits will be a sample of careers in which people work with matter, machines, and energy. You may work by yourself or in a group.

Machine Designer

Imagine that you are a designer of roller coaster rides. You have to find a way to make cars go faster down an inclined plane. Experiment with a toy car on a model of a roller coaster. Demonstrate how you changed the car to make it go faster.

Environmental Officer

Imagine that you are a safety officer in your community. Decide which sounds in your classroom or town could be harmful. Write some new rules that would help get rid of these noises.

Life Scientist

Suppose you are an expert on animals and how they hear. Make a poster called "Unusual Ears." Draw or cut out pictures of animals with unusual ways of receiving sound. Write a sentence for each picture explaining how that animal hears. Make a second poster called "Unusual Voices." Show animals that use different parts of their bodies to make sounds.

Cook

Imagine that you are a cook in a restaurant. Describe how you prepare some dishes. Tell which steps involve physical changes and which involve chemical changes.

Construction Worker

Suppose that you are a construction worker showing others how a lever helps you lift a heavy object. Use a metric ruler as a lever. Tape three pencils together to make a fulcrum, and use pennies to model the forces. Show how a small number of pennies can lift a larger number of pennies. Move the pennies and the fulcrum and observe what happens.

Making Tables and Writing a Description

You can use a table to organize ideas before you write. A table lists information about a topic or subject in columns and rows. The table below lists some information about the forms of energy you studied in Chapter 3.

Some Simple Machines

Type of Machine	How It Helps Work
Lever	
Inclined Plane	
Pulley	
Wedge	

The Different Forms of Energy

Form of Energy	Description
Heat	Heat moves from a warm object to a cooler one.
Light	Light travels in straight lines and can be reflected.
Electricity	All matter has electric charges. Electricity can flow as electric current in an electric circuit.

Make a Table

On a separate sheet of paper, make a table like the one shown above at the right. Include enough rows in your table for all of the six types of simple machines. Use what you learned in Chapter 2 to fill in your table.

Write a Description

Imagine you have to move a heavy carton of books upstairs. Use the data in your table to decide which machines to use. Write a paragraph to describe your choice.

Remember to:

1. **Prewrite** Organize your thoughts before you write.

2. **Draft** Write your description.

3. **Revise** Share your work and then make changes.

4. **Edit** Proofread for mistakes and fix them.

5. **Publish** Share your description with your class.

Unit C
Earth Science

Science and Technology
In Your World!

Boring, Boring, Boring!

What has diamond or metal teeth, is 28 feet tall, and chews rock? It's a "TBM" or Tunnel-Boring Machine. These monster machines drilled a tunnel in the rock under the waters of the English Channel. Now trains carrying people, cars, and trucks speed through the "Chunnel," a tunnel that connects France with England. You'll learn more about the earth's surface and how people affect it in **Chapter 1 Changes in the Earth's Surface.**

Signals Through Glass

You see through it. It's made of a melted mixture of sand, chalk, and soda. What is it? It's glass. Sometimes glass is spun into tiny threads that carry light. These threads help doctors see inside the human body. They also help other scientists look inside places. You'll learn more about products made from the earth's mineral resources in **Chapter 2 Materials of the Earth.**

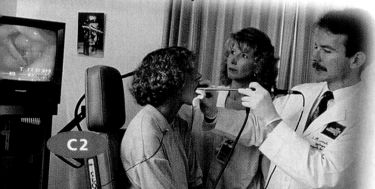

Rain Seen from a Plane!

To study clouds, sometimes scientists need a closer look. This plane flies into clouds loaded with cloud-measuring instruments. For starters, it has a camera that takes video shots of very tiny bits of matter in the clouds. It also has tools that count water droplets and measure water amounts. Other tools measure the size of water droplets, as well as how they form, grow, and join other droplets. You will learn more about clouds and the clues they give about weather in **Chapter 4 Clouds and Storms.**

Robot Sends Pictures from Space!

Did you know that robots have explored other planets? This robot roamed around Mars. The six-wheeled microrover was called *Sojourner*. On Mars, *Sojourner* wandered over rocks called "Shark," "Wedge," and "Sea Cucumber." It sent radio signals and pictures back to scientists on Earth. You will read more about how scientists study the solar system in **Chapter 3 The Sun, Planets, and Moon.**

It's Alive?

Well, no, the earth isn't alive, but it's always changing. Look outside. You probably don't see the earth changing. Most changes happen very, very slowly. It took a lot of changes over millions of years to make Bryce Canyon look the way it does now.

Chapter 1
Changes in the Earth's Surface

Inquiring about Changes in the Earth's Surface

Lesson 1
How Do Volcanoes and Earthquakes Change the Earth?

How do volcanoes form?

How do volcanoes change the earth?

How do earthquakes change the earth?

How do scientists study volcanoes and earthquakes?

How can you be safe during earthquakes?

Lesson 2
What Landforms Are on the Earth's Surface?

Where are some landforms?

What are some kinds of landforms?

Lesson 3
How Do Water and Wind Change the Earth's Surface?

How does weathering change rocks?

How does erosion change the earth?

Lesson 4
How Can Living Things Affect the Earth's Surface?

How do plants and animals change the earth?

How do people change the earth?

Copy the chapter graphic organizer onto your own paper. This organizer shows you what the whole chapter is all about. As you read the lessons and do the activities, look for answers to the questions and write them on your organizer.

Exploring Rocks Formed from Molten Rock

Process Skills

- observing
- classifying
- communicating

Materials

- piece of paper
- granite
- obsidian
- pumice
- hand lens
- balance

Explore

1. Write the words *Granite, Obsidian,* and *Pumice* on a piece of paper. These rocks formed when molten rock cooled and hardened. Place each rock next to its name.

2. Use a hand lens to **observe** the rocks. Describe the rocks. List as many properties as you can.

3. Put one rock in one balance pan. Put another rock in the other pan. Compare the masses of all three rocks. **Classify** the rocks by arranging them in order of least mass to greatest mass.

Reflect

1. Pumice has less mass than granite or obsidian of similar size. What do you observe about pumice that may explain this?

2. Communicate. Discuss your observations with the class.

? **Inquire Further**

How are these rocks similar to or different from rocks in your area? Develop a plan to answer this or other questions you may have.

Identifying the Main Idea

As you read for science, it is important for you to find the main ideas for each lesson.

Sometimes these main ideas are stated directly. Look at "What's the Big Idea?" on page C8. The five items in this list are the stated main ideas for this first lesson. Every lesson in your book lists the stated main ideas in this same place. As you read the lesson, however, the main ideas might be paraphrased or written in a slightly different way.

▼ *How can you find the main idea about what's happening here?*

Example

The first stated main idea for Lesson 1 is "how volcanoes form." In the lesson it has been rewritten as "Magma pushes upward and erupts from cracks in the earth's surface."

Use the following chart for the rest of Lesson 1. Make a copy of the chart on your own paper. Write the last four main ideas from "What's the Big Idea?" in the first column. As you read the lesson, decide whether these main ideas are stated or paraphrased. Write the main idea sentences in the correct column.

Main Ideas	Stated	Paraphrased

Talk About It!

1. Where can you find the main ideas for each lesson in your book?

2. What two types of main ideas will you find in your book?

You will learn:

- how volcanoes form.
- how volcanoes change the earth.
- how earthquakes change the earth.
- how scientists study volcanoes and earthquakes.
- what to do during an earthquake.

Glossary

volcano (vol kā′nō), a type of mountain that has an opening at the top through which lava, ash, or other types of volcanic rock flow

Kilauea is in Hawaii. Hot, melted rock and ash are erupting and are covering the mountain. ▶

Lesson 1

How Do Volcanoes and Earthquakes Change the Earth?

You shake a can of whipped cream. You push on the nozzle. **WHOOSH!** Whipped cream explodes out of the opening. Now you know what can happen if forces build up. A volcano is a little like a whipped cream can.

How Volcanoes Form

Hot, melted rock and gases are bursting out of the volcano in the picture. A **volcano** is a special type of mountain with an opening, or vent, at its top. Thousands of years ago, this volcano did not exist. So what happened to form this volcano?

To understand volcanoes, you need to know something about the inside of the earth. The inside of the earth is very hot. It is so hot that some rocks melt and become a thick, flowing material called **magma.** Magma gathers far underground in places called magma chambers. Magma can build up in a magma chamber for hundreds or thousands of years.

Powerful forces build up in a magma chamber. Eventually, the magma pushes upward and then erupts from cracks in the earth's surface. **Erupt** means to burst out. An eruption can be very strong or rather gentle. Magma that erupts out onto the earth's surface is called **lava.** The lava is made of melted rock. The diagram shows some things that happen when a volcano erupts.

Glossary

erupt (i rupt′), to burst out

lava (lä′və), hot, melted rock that comes out of a volcano

magma (mag′mə), hot, melted rock and gases deep inside the earth

4 *Lava builds up outside the volcano. It cools and hardens. A mountain forms.*

3 *Lava erupts from the volcano.*

2 *Magma pushes upward through cracks in the earth.*

1 *Magma rises from deep inside the earth. It gathers in a magma chamber.*

How Volcanoes Change the Earth's Surface

Eventually, the hot lava cools into solid rock. You can probably guess what happens as more and more lava builds up on the earth's surface—a mountain or volcano forms. A volcano might erupt many times. Each time a volcano erupts, more lava spreads over the volcano. The volcano gets taller and taller.

Volcanoes can change the earth's surface quickly. Sometimes, a volcano forms under the ocean. If the volcano gets tall enough, it will reach the surface of the water. Some islands are actually formed from volcanoes that reached the ocean's surface. Hawaii and Japan are groups of islands that formed from volcanoes.

The island of Surtsey is near Iceland. It is the top of a volcano. It formed when the volcano erupted under the sea. ▼

How Earthquakes Change the Earth's Surface

Like volcanoes, earthquakes can change the earth's surface quickly. An **earthquake** is a shaking or sliding of parts of the earth's surface. Layers of huge blocks of rock lie far beneath the earth's surface. Earthquakes happen when some of these blocks move.

The surface of the earth has many large cracks. When the blocks of rock move, new cracks can form. The moving blocks of rock also cause the land above them to move. The land may move up, down, or sideways. How has an earthquake changed the surface in the picture?

What causes the blocks of rock to move? In certain places deep inside the earth, the blocks keep pushing against each other. For many years, the blocks do not move. The pushing keeps going on. Finally, one or both of the blocks move. Think about putting two pieces of cardboard side by side on your desk. If you push them together harder and harder, the pieces would finally bend or move.

Glossary

earthquake (ėrth′kwāk′), a shaking or sliding of the surface of the earth

◀ *An earthquake happened here. Imagine how the road looked before the earthquake.*

How Scientists Study Volcanoes and Earthquakes

Most volcanoes and earthquakes happen in the same general areas of the world. Remember that earthquakes occur where huge, moving blocks of the earth's surface are pushing together. These moving blocks can cause deep cracks to form in the earth's surface. Remember also that magma erupts from cracks in the earth's surface. Earthquakes often happen during or just before the eruption of a volcano.

Scientists try to predict earthquakes. They use several different instruments to pick up very slight movements under the earth. These movements also happen before volcanoes erupt. Satellites out in space can also pick up warning signs that an earthquake might happen or a volcano might erupt.

This scientist is called a volcanologist. She studies volcanoes. She is taking samples of lava from a volcano in Hawaii. The lava will give her information about the inside of the earth. Why do you think she is wearing a special suit? ▼

Safety During Earthquakes

Earthquakes can be very dangerous. Sometimes they happen without any warning. If you live in an area that has earthquakes, you can help protect yourself if you know what to do.

You might have an earthquake kit in your home. It should have flashlights, batteries, a fire extinguisher, a radio, a first-aid kit, water, canned food, and warm clothes.

You should know how to Duck, Cover, and Hold, as in the picture. Remember to stay away from windows or furniture that could fall over. Stay away from trees, signs, windows, and electric wires. When you are in school, follow your teacher's instructions.

Duck, Cover, and Hold

Duck, or drop down to the floor. Take cover under a desk or strong table. Hold onto the table and move with it. Stay in this position until the ground stops shaking and it is safe to move. ▼

Lesson 1 Review

1. How does a volcano erupt?
2. How do volcanoes change the earth?
3. How do earthquakes change the earth?
4. How do scientists study earthquakes and volcanoes?
5. What are some ways to stay safe during an earthquake?
6. **Main Idea**
   Read the material on page C 10. What is the main idea of this material?

What's the Big Idea?

You will learn:

- about some landforms in North America.
- to describe landforms on the earth's surface.

Glossary

landform
(land′fôrm′), a shape on the earth's surface

Lesson 2

What Landforms Are on the Earth's Surface?

Pack your bags! You're going to travel across the United States! You will see different kinds of land features. Some parts of the earth will look like where you live. Some parts will look very different.

Landforms in North America

The surface of the earth has many different shapes. Each kind of shape is a **landform.** Three important kinds of landforms are mountains, plains, and plateaus.

◀ *Rocky Mountains*

Colorado Plateau ▶

Recall that volcanoes are mountains that are built quickly. Other mountains are formed more slowly by forces deep inside the earth. The Rocky Mountains and the Appalachian Mountains were built by pushing forces over millions of years. The forces are still building the mountains.

Near the center of the United States is the area called the Great Plains. A **plain** is a large, flat area of land. The United States has two main areas of plateaus. A **plateau** is a large, flat area that is high.

Glossary

plain (plān), a large, flat area of land

plateau (pla tō′), a large, flat area of land that is high

Glossary

◀ Great Plains

Appalachian Mountains ▶

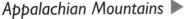

◀ Ozark Plateau

Kinds of Landforms

Glacier

A glacier is a huge area of moving ice. Glaciers form in high mountains and in other cold parts of the world. They also form in the area near the north and south poles. A glacier moves very slowly down a mountain. It can move through a valley or over a wide area of land.

There are many more kinds of landforms than just mountains, plains, and plateaus. All landforms are alike in one way. They all are shaped by some type of force. You learned that volcanoes and other mountains are formed by forces deep within the earth.

Valley

A valley is a low area between mountains or hills.

Bay

A bay is a part of an ocean or a lake. The shore of a bay goes in toward the land.

Plateau

A plateau is like a plain, but a plateau is high.

Cliff

A cliff is a very steep wall of rock or soil. Many cliffs formed because water or glaciers pushed material from the sides of hills or mountains. This kind of change took thousands of years.

Ocean

The ocean is all the salt water that covers most of the earth's surface. The major parts of the ocean are also known as oceans: Atlantic Ocean, Pacific Ocean, Indian Ocean, and Arctic Ocean.

Other forces act on the earth's surface. Moving water and wind apply forces to materials. They can work to shape a landform. Notice how water and wind may have affected the landforms on these two pages.

Mountain
A mountain is a very high place on the earth's surface. Mountains usually have pointed tops.

River
A river is a stream of water. The moving water in a river usually flows toward a lake or the ocean.

Lake
A lake is a large body of water that is almost totally surrounded by land. A lake is larger than a pond.

Plain
A plain is a large, mostly flat area of land.

Hill
A hill is a high place on the earth's surface. Hills are smaller and not as tall as mountains. Hills have rounded tops.

Coast
A coast is a place where the land meets the ocean. Each area of coast has a special look. Some coasts are sandy beaches and others are rocky cliffs. Wind and moving water are always changing coasts.

Lesson 2 Review

1. Where are the major landforms in North America?

2. Describe some landforms and some areas of water on the earth's surface.

3. Main Idea
Read the paragraph at the top of page C16. Identify the sentence that describes how landforms are alike.

Making a Model of a Landform

Process Skills

- making and using models
- predicting
- observing
- inferring

Materials

- safety goggles
- plastic cup
- damp sand
- foil pan
- pebbles
- thick book
- dropper
- water in plastic cup

Getting Ready

In this activity you will make a model of a landform and observe the effects of moving water on land.

Follow This Procedure

❶ Make a chart like the one shown. Use your chart to record your predictions and observations.

	Predictions	Observations
Drops of water		
Poured water		

❷ Put on your safety goggles. Put a cupful of damp sand at one end of the pan. Make a **model** of a mountain out of the sand.

❸ Cover the mountain with pebbles. Pat them gently so they stay in place.

❹ Place the book under the end of the pan that has the mountain (Photo A).

❺ Make a **prediction.** What would happen if you dropped single drops of water on the mountain? Record and explain your predictions.

❻ Fill the dropper with water. Slowly drop 5 drops of water on the mountain (Photo B). Record your **observations.**

 Safety Note Wipe up any spills immediately.

Photo A

Photo B

 Repeat steps 5 and 6, but this time slowly pour the water from the cup over a different place on the mountain. Record your observations.

Self-Monitoring
Have I correctly completed all the steps?

Interpret Your Results

1. How did the drops of water affect the sand and pebbles? How did the poured water affect the sand and pebbles?

2. Make an **inference** based on your observations. How can moving water affect landforms?

Inquire Further

What landforms can you find in your community or state? Where can you find examples of how water affects the land? Develop a plan to answer these or other questions you may have.

Self-Assessment

- I followed instructions to make a **model** of a landform.
- I recorded my **predictions.**
- I **observed** the effects of moving water.
- I recorded my observations.
- I made an **inference** about the effects of moving water on a landform.

You will learn:

- how weathering changes rocks.
- how erosion changes the earth's surface.

Glossary

weathering
(weth′ər ing), the breaking apart and changing of rocks

Lesson 3

How Do Water and Wind Change the Earth's Surface?

Snow, ice, water, a plant—they seem like pretty gentle things. Can you believe that they all can break a rock? They can—but only after a very long time.

How Weathering Breaks Rocks

Physical Science   Changes are happening to landforms all the time. Some changes happen very quickly. The changes that result from erupting volcanoes are examples of changes that happen quickly. Other changes are slow and may go on for hundreds, thousands, or even millions of years. All surfaces on the earth can change. Even a material as hard as rock can change.

Weathering is usually a slow change. **Weathering** is the breaking down and changing of rocks. Weathering can be caused by water, changing temperatures, and living things.

◀ A pothole starts to form when water gets into a tiny crack in the street surface. When the weather becomes cold, the water freezes. When it becomes warmer, the ice melts. All the freezing and melting put pressure on the pavement. The small crack grows into a larger crack. When cars and trucks pass over it, the large crack can grow into a round hole.

After many years of weathering, a huge boulder might crack and crumble. The tops and sides of mountains also can be changed by weathering.

Water causes weathering in a few ways. Water can get into cracks in rocks. When water freezes, it expands, or grows larger. It pushes on the sides of the cracks. Eventually, the rocks may break apart. Water can also dissolve some parts of rocks. The dissolved rocks might be washed away. In some places, caves may form where water dissolved away rock.

Plants can also break down rocks. Plant roots can grow down inside cracks in rocks. As the roots grow larger, they can break apart the rocks. The roots of a large plant, such as a tree, can even break a huge boulder.

Freezing and melting are breaking down the rock in the cliff. The snow will melt, and the water will get into cracks in the rocks. Then the water will freeze. After some years, cracks in the rock will grow. Pieces of rock might fall off the cliff. ▶

Glossary

erosion (i rō′zhən), the carrying away of weathered rocks or soils by water, wind, or other causes

glacier (glā′shər), a huge amount of moving ice

Rivers

This canyon in Utah was formed by the moving water of the Green River. Over thousands of years, the river water carried away bits of weathered rock. At some places, the force of the moving water was very strong. The bits of rock and sand in the water wore away more of the river walls. Erosion is making the canyon deeper and deeper. ▼

How Erosion Changes the Earth's Surface

What happens to rocks that have broken apart because of weathering? Some pieces stay where they are. Other pieces get carried away to other places. Each year, wind, water, and gravity move weathered rocks and soils. The carrying away of these materials is **erosion.**

Erosion changes the earth's surface everywhere. Most of the time, these changes happen slowly. Sometimes, though, erosion can happen quickly. The pictures show some ways that erosion happens.

▲ Glaciers

*The area that looks like a large road is a glacier. A **glacier** is a huge amount of moving ice that forms over many years. It is made of layers of snow. Glaciers move slowly. They carry rocks with them as they move. As the glacier melts, the rocks stay on the ground.*

◀ Wind

Wind can carry away only small bits of material, such as dust or sand. Wind erosion happens most often in dry, sandy places. Strong winds can carry sand over long distances. When the wind slows down, the sand drops to the ground. Wind and sand can also shape objects like this pillar. The blowing sand kept hitting against a pile of rock. After many years, some of the rock was scraped away.

▲ Gravity

You have learned that gravity is a force that pulls objects down. Gravity can also cause erosion. It causes rocks and soils to move down slopes. Sometimes they move quickly, as in this mud slide. A great amount of rain fell here. The wet, heavy soil moved quickly down the slope.

Lesson 3 Review

1. What does weathering do to rocks?

2. Describe some ways that erosion happens.

3. **Main Idea**
   Look at the captions for the pictures on pages C20 and C21. Find one sentence in each caption that tells how weathering is alike in the pavement and in the cliff.

You will learn:

- how plants and animals can change the earth.
- how people can change the earth's surface.

Lesson 4

How Can Living Things Affect the Earth's Surface?

It's a great day for the beach! You pack a lunch and put on sunscreen. At the beach, you make a fantastic sand castle. You have changed the earth's surface. Don't worry. The wind and waves will change it again.

Tree roots were strong enough to break this sidewalk. What would they do to a rock? ▼

Ways that Plants and Animals Can Change the Earth

Life Science

Recall that the roots of plants can cause weathering. The roots of trees can be especially strong. Look what they have done to the sidewalk. As the tree grew, the roots grew too. They became strong enough to crack the sidewalk. Roots can crack and move big boulders too.

Trees and other plants also help the earth's surface. They protect the soil from erosion by wind and rain. Trees act as a shield against wind. Their roots go deep into the soil. Roots help hold soil in place even when there is a great deal of rain.

Plants are especially helpful on hills or other slopes. Here, their roots help hold the soil and protect it against erosion by gravity.

Animals change the earth's surface all the time. Some animals live deep in the soil. They change the surface of the earth as they dig through the soil. Earthworms, ants, and other small animals make tiny tunnels in soil. The tunnels allow water and air to mix into the soil. The wastes that these animals leave also add nutrients to the soil.

Other animals, such as groundhogs and these prairie dogs, also live in the ground. How do their homes change the earth's surface?

A prairie dog town ▼

Building Roads

When people build roads, they dig into the earth. When the roads go through or around mountains, erosion can be a special danger. What kind of erosion should people worry about on this road in Asia? ▼

Ways That People Change the Earth's Surface

People change the earth's surface when they dig into it, build on it, and plant in it. People do some things that cause erosion. They also do things that prevent erosion. The pictures on these two pages show some ways that people cause and prevent erosion.

Wind Fences

People put fences like these on beaches. Water might cause erosion here. Wind also blows away a great deal of sand. The fences shield the sand. They keep some of it from being blown far away. ▼

▲ Farming

Farmers change the earth's surface in order to raise food crops. Farmers need to know the best way to plant crops in order to prevent erosion. They also must know which kinds of crops will help keep the soil in place. Sometimes, though, floods or other events happen. Then soil might be washed away, as on this farm.

▲ Terrace Farming

In wet, hilly areas, farms might look like this one. These steps, or terraces, are cut into a hillside in Asia. This kind of farming prevents erosion by water and by gravity. The fields are flat across the hill, instead of being on a slope. Soil would not be as likely to fall or be washed downhill.

Lesson 4 Review

1. What are some ways that plants and animals change the earth's surface?

2. How do people cause or prevent erosion?

3. **Cause and Effect**
   Make a chart that lists the causes and effects of ways that living things change the earth's surface.

Chapter 1 Review

Chapter Main Ideas

Lesson 1

• Volcanoes form when lava erupts at the earth's surface and hardens.

• Volcanoes change the earth's surface by building mountains and islands.

• Earthquakes move the earth's surface and cause cracks to form.

• Scientists examine lava, and they use special instruments to study volcanoes and earthquakes.

• To be safe during an earthquake, follow the teacher's instructions, have a safety kit, and do Duck, Cover, and Hold.

Lesson 2

• Mountains, plains, and plateaus are important landforms.

• Landforms and bodies of water include cliffs, hills, coasts, valleys, lakes, rivers, bays, and oceans.

Lesson 3

• Water, changes of temperature, and living things cause weathering.

• Erosion happens when wind, water, gravity, and other things move rocks and soils.

Lesson 4

• Plant roots change the earth by breaking rocks and by helping to prevent erosion; animals change the earth by digging holes and tunnels.

• People change the earth's surface by digging and by preventing erosion.

Reviewing Science Words and Concepts

Write the letter of the word or phrase that best completes each sentence.

a. earthquake   g. magma

b. erosion   h. plain

c. erupt   i. plateau

d. glacier   j. volcano

e. landform   k. weathering

f. lava

1. A type of mountain that forms from hardened lava is a ___.

2. Volcanoes ___ when magma rises up and comes to the earth's surface.

3. Melted rock that is deep within the earth is ___.

4. A high, flat area of land is a ___.

5. A ___ is a shape on the earth's surface.

6. Rocks that have been changed or broken have undergone ___.

7. Hot material that comes out of a volcano is called ___.

8. A ___ is a large, flat landform.

9. The surface of the earth might move during an ___.

10. A ___ is a huge amount of moving ice.

11. When earth materials are moved away from a place, ___ occurs.

Explaining Science

Draw and label a diagram or write a paragraph to answer these questions.

1. How is the way that an earthquake changes the land different from the way that erosion changes the land?

2. How are hills different from mountains and plateaus?

3. How can water cause weathering of rocks?

4. How can tree roots cause weathering of rocks?

Using Skills

1. Look at page C21. What is the **main idea** about how weathering can affect a mountain?

2. Sand and soil are always moving across the land because of erosion. How can erosion affect sandy beaches? **Communicate** your thoughts by writing a paragraph.

3. Look at the diagram of landforms on pages C16 and C17. Which of these landforms do you **observe** in your area?

Critical Thinking

1. **Make a generalization** to explain how erosion can be both helpful and harmful.

2. Many plants grow well in soil in which animals live or dig. What would you **infer** about the soil in these places?

3. **Predict** where more erosion will happen—in a thick forest or on a bare hillside.

Smart Art!

You can use all sorts of materials to make a work of art—clay, wood, paint, or what some people call trash. Your imagination can help you create something beautiful from anything you find. Look at what this artist made! What can you find in your recycling bin?

Chapter 2
Materials of the Earth

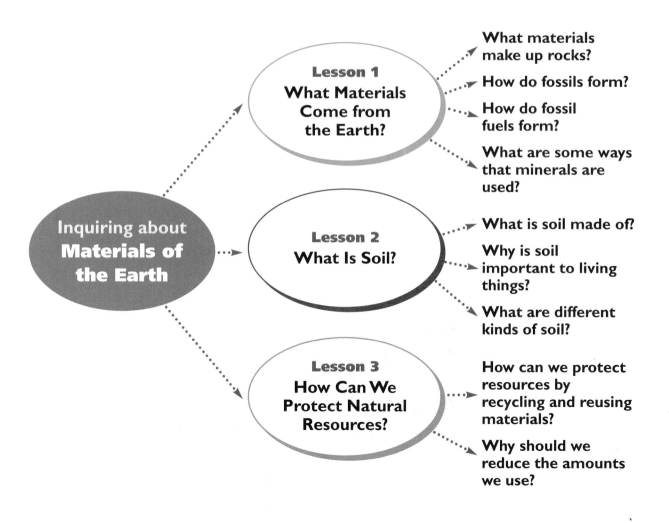

Lesson 1
What Materials Come from the Earth?

What materials make up rocks?

How do fossils form?

How do fossil fuels form?

What are some ways that minerals are used?

Inquiring about Materials of the Earth

Lesson 2
What Is Soil?

What is soil made of?

Why is soil important to living things?

What are different kinds of soil?

Lesson 3
How Can We Protect Natural Resources?

How can we protect resources by recycling and reusing materials?

Why should we reduce the amounts we use?

Copy the chapter graphic organizer onto your own paper. This organizer shows you what the whole chapter is all about. As you read the lessons and do the activities, look for answers to the questions and write them on your organizer.

Exploring Properties of Rocks

Process Skills

- observing
- classifying
- communicating

Materials

- small rocks of different kinds
- hand lens
- paper

Explore

1 **Observe** the rocks carefully. Think of a way to put the rocks into groups. **Classify** the rocks according to your grouping system. Record the properties of each group.

2 Choose one rock. Observe it carefully with the hand lens. How is it like the other rocks in its group? How is it different from the others? Record all of your observations on a piece of paper.

3 **Communicate.** Exchange papers with one of your classmates. Have your classmate pick out the rock you described. Then try to pick out the rock that your classmate described.

Reflect

1. Which properties did you use to classify the rocks?

2. Which properties did you use to describe your rock?

? Inquire Further

How can you classify rocks found in your area? Develop a plan to answer this or other questions you may have.

Exploring Multiplying Tens

What can be done with the millions of plastic bottles that are recycled each year? Some people have found a way to make fuzzy cloth out of them. Did you ever think plastic bottles could keep you warm?

▼ *The bottle will be chopped into flakes. The flakes will be melted and then spun into yarn.*

Work Together

Find each product. You can skip count by tens to find the total.

1. One pound of cloth can be made from 10 recycled plastic bottles. How many plastic bottles are needed to make 4 pounds of cloth?

2. If it takes 20 plastic bottles to make a sweatshirt, how many plastic bottles are needed for 3 sweatshirts?

3. Write your own problem about recycled plastic bottles. Then solve.

Talk About It!

How did you find the total number of bottles needed for 4 pounds of cloth?

You will learn:

- how minerals make up rocks.
- how fossils form.
- how fossil fuels form.
- how minerals are used for making objects.

Glossary

mineral (min′ər əl), material that forms in the earth from nonliving matter

Lesson 1

What Materials Come from the Earth?

It's your turn to help make dinner. You take out some pots and pans. You put dishes and glasses on the table. Do you realize that you're using materials that came from the earth?

Rocks and Minerals

In the last chapter, you read that rocks are underneath landforms. There are many different kinds of rocks. Some kinds of rocks are used for building. Large blocks of rocks for building are dug out of the earth. A quarry like the one shown is a place where rocks are dug from the earth.

All rocks are made of substances called minerals. A **mineral** forms in the earth from nonliving matter. Different kinds of rocks are made of different combinations of minerals.

◀ *Blocks of stone are being dug from the earth. These blocks will be used for making buildings.*

There are over 4,000 different minerals. Each mineral has special properties. Some minerals are shiny and some are dull. Some minerals are hard and others are soft. Each has a color. A rock can be made of one or more minerals. You might see several colors in a rock. Each color is a different mineral. What are the properties of the rose quartz shown below?

Rocks form in three main ways. Look at the picture of granite. It starts as melted material deep within the earth. The material moves toward the earth's surface and cools. This material turns into rock even while it's still below the earth's surface.

Rocks such as sandstone form in another way. Notice its layers. These rocks form from layers of sand, bits of rocks, soils, and sometimes shells. Over millions of years, the layers all become cemented together.

A third kind of rock forms as a result of changes in the other two types. Very high heat and great pressure inside the earth can change rocks. The changes occur over millions of years. Gneiss (pronounced "nice") formed in this way. It is shown at the right.

▲ *Granite is a rock that formed from cooled material inside the earth. How many minerals can you see in it?*

▲ *Sandstone is a rock that formed as layers of sand got pressed together.*

▲ *Gneiss formed when great pressure on rocks changed the forms of the minerals within it.*

◀ *This is a mineral called rose quartz.*

Glossary

fossil (fos′əl), the hardened parts or mark left by an organism that lived a long time ago

▲ *This fossil is called* Archaeopteryx (är′kē op′tər iks). *It was found in Germany in 1861. It lived about 150 million years ago. Most scientists think that it was a type of bird.*

Fossils

As some rocks form, parts of organisms may remain in the material that forms the rock. You might sometimes find a rock that has a mark or a part of an organism that died long ago. Notice such a rock below. It has a fossil of an ammonite in it.

Life Science

A **fossil** is a hard material with traces of an organism that lived a long time ago. A fossil might be a whole organism or a part of one. A fossil might also be a footprint that an animal left in mud or sand. Over many years, the mud became hard and turned into rock.

Sometimes parts of a dead organism decay in mud or sand. Over many years, the mud turns to rock. Then just a mark remains where the organism was pressed into the rock. The mark has the same shape as the organism.

Animals and plants that lived millions of years ago left many fossils in rocks. Scientists study these fossils to learn what living things looked like so long ago. The fossil above gives clues about certain birds that became extinct millions of years ago.

◄ *Ammonites were animals that lived in the sea. They became extinct around the same time as dinosaurs.*

Fuels

Some rocks have materials that we use as fuels. A **fuel** is a material that can be burned to produce useful heat. We use fuels to warm our homes and to cook. Power plants like the one shown burn fuel to make electricity.

Some fuels are called fossil fuels. They usually do not have actual fossils that we can see. Fossil fuels got their name because they formed from organisms that lived hundreds of millions of years ago. Coal and oil are fossil fuels.

Most coal formed from plants that grew in swampy areas. The plants died and fell to the bottom of the swamp. Over millions of years, many layers of plant material formed. The swamps dried up, and the layers of plant material were pressed into rock like the piece of coal shown at the right.

Many scientists think that oil formed from microscopic organisms that lived in the ocean. Their remains piled up on the ocean floor. Sand and other materials covered these remains. Over millions of years, the layers of remains went through a chemical change and became oil. Parts of the ocean dried up, but oil stayed buried in rocks underground. Pumps like the one shown are used to get the oil.

Glossary

fuel (fyu′əl), a material that can be burned to produce useful heat

Power plants burn fuel to produce electricity. ▼

Coal is a fuel that formed mostly from plant material. Coal is dug from the earth at coal mines. ▼

At an oil well a pump brings the liquid fuel up from underground. ▶

C37

natural resource
(nach′ər əl rē′sôrs),
a material that comes
from the earth and can
be used by living things

Making Things from Minerals

Rocks and the minerals they contain are natural resources. A **natural resource** is a material that comes from the earth and can be used by living things. You learned that some rocks are used in buildings. Other rocks can be used as fuel. The minerals that make up some rocks can be used to make many simple everyday materials. Glass and metals are made from minerals. Follow the steps in making glass objects from sand.

1 Materials for Making Glass

A large amount of sand is mixed with smaller amounts of lime (from limestone) and soda (from another mineral). Bits of recycled glass might also be added.

2 Heating the Mixture

The sand and other materials are fed into a furnace. The furnace is very hot—about 1400°C. At this temperature, the materials melt into a liquid.

3 Making Bottles

A lump of melted glass is dropped into a mold. Air is blown into the glass. The glass takes the shape of the mold and is hollow inside. The bottle comes out of the mold and is left to cool and become hard.

Some rocks—called **ores**—have large amounts of useful minerals. People have learned how to separate these minerals from the ores. Iron, copper, and aluminum are some metals that come from ores.

Steel is a very important metal that is made from iron ore. Steel is used for making nails, cars, and almost every machine you might think of. The pictures show some steps in making steel.

Glossary

ore (ore), rock that has a large amount of useful minerals

1 Ore from the Mine
Iron ore is dug from huge open-pit mines. Power shovels scoop out the ore.

2 Materials for Making Steel
Steel is made from iron ore such as this. The ore is combined with limestone and a form of coal.

3 In the Furnace
The materials are put into a very hot furnace. As they burn, a chemical change happens. Iron is separated from the ore. Melted iron collects at the bottom of the furnace. It is poured out and moved to another furnace to be made into steel.

Lesson 1 Review

1. What are rocks made of?
2. How do fossils form in rocks?
3. Why are some fuels called fossil fuels?
4. What are some materials that are made from natural resources?
5. **Sequencing**
   Place the following sentences in the correct sequence:
   a. Plant material is changed into coal.
   b. Swamps dry up.
   c. Dead plants fall to the bottom of the swamp.
   d. Layers of plant material become pressed together.

What Is Soil?

You will learn:

- what soil is made of.
- about soil and living things.
- to describe different kinds of soil.

Think of all the ways people might use this tree when it grows. ▼

Have you ever gone to a beach? When you were there, you might have played in the sand and made sand castles. Did you ever wonder why so few plants grow in sand?

What Soil Is Made Of

Life Science

You read in the last lesson that rocks are natural resources. Soil is a natural resource too. The students in the pictures are using soil as farmers all over the world use it. People use soil to grow the plants that we all need.

Would you be surprised to learn that soils form from rocks? Recall that water and living things cause weathering of rocks. As they weather, rocks break apart into smaller and smaller pieces. Over many years, some of the rock minerals dissolve and get washed away. Other minerals help soil form.

◀ *These vegetable plants will be used as food.*

C40

Soil is made of more than bits of rock. It also has water, air, and material that once was alive. When organisms die, they **decay**. To decay means to break down or rot. The decayed material also becomes part of the soil.

The decayed remains of plants and animals in soil is called **humus**. Humus is usually dark brown or black. Humus adds nutrients to soil. A **nutrient** is a material that plants and animals need to live and grow. The pictures show what happens when soil forms.

Glossary

Glossary

decay (di kā´), to slowly break down or rot

humus (hyü´məs), decayed organisms in soil

nutrient (nü´trē ənt), a material that plants and animals need to live and grow

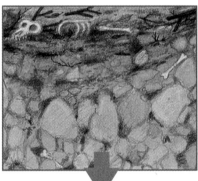

1 Rocks begin to weather.

Recall that weathering can take thousands of years. The rocks start to break down into smaller and smaller pieces. Parts of dead plants and animals are mixed with the rocks.

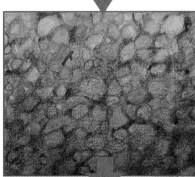

2 Rocks continue to break down.

The rocks break down into smaller pieces. The dead organisms decay. They become humus.

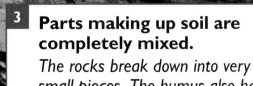

3 Parts making up soil are completely mixed.

The rocks break down into very small pieces. The humus also has completely broken down so that you cannot see plant or animal parts. The rock pieces and humus become mixed. Soil has formed.

Soil and Living Things

All living things need the earth's soil. Plants need the minerals and other nutrients in soil to live and grow. The animals that eat plants depend on soil. Some animals make their homes in soil. Plants and animals take nutrients from the soil. They also add nutrients to it. The picture shows some of the activities of life underground. Notice the different layers of the soil.

Leaves

Leaves and other plant parts fall to the soil. They decay there and help form humus.

Ants and Other Insects

Many insects dig tunnels in the soil. They make nests and lay their eggs here. Animals also add nutrients to soil. When they die underground, their bodies decay and over time become humus.

Roots of Plants

Roots push deep into the soil. They take water and nutrients from the soil. Roots also hold the plant in the soil. Roots help the soil because they hold the soil in place. Soil erosion does not happen quickly where many plants are growing. Roots of some plants also add nutrients to the soil.

Top Soil Layers

Roots and animals are in the top layers of soil. Humus is also in the top soil layers. Some small pieces of rock might be in the top layers too.

Rocky Layers

Beneath the top soil layers are lower layers of soil that do not have much humus. Under these soil layers are layers of rocks. Higher up, the rocks might be in pieces. Solid rock lies beneath layers of broken rocks.

Earthworms

Earthworms and some spiders make their homes underground. They dig tunnels in the soil. The tunnels allow air, water, and nutrients to pass easily through the soil. The tunnels also make it easier for the roots of plants to grow and get these important materials.

C43

Glossary

Kinds of Soil

Think about soil you have seen and handled. Could you dig easily into it? Did it have big chunks or tiny pieces? You might have noticed that soil looks different in different places. It can have different colors. It also can have pieces of different sizes. Soil can feel very hard or quite soft. It might feel dry or be wet and squishy.

The pictures show three kinds of soil. Each kind is a different color. In each kind, the pieces are a similar size. Each holds water in a special way. Each kind of soil has certain minerals. Some plants can grow in each kind of soil. However, all plants cannot grow in all kinds of soil.

Clay soil has tiny pieces. It feels smooth because the pieces are very close together. Clay soil has many nutrients. However, many plants cannot grow well in clay soil. Their roots cannot push through the soil because the soil pieces are packed together so closely. Water cannot pass through clay soil very well. Parts of the United States along the Mississippi River have clay soil. Some bushes and shrubs grow there. ▶

Loam is a mixture of clay, silt, sand, and humus. Loam holds water well and has many nutrients. It is lighter and looser than clay soil. Most plants grow well in loam. This wheat farm in the Great Plains has loam soil. ▼

Glossary

Glossary

loam (lōm), good planting soil that is a mixture of clay, silt, sand, and humus

sandy (san′dē) **soil,** loose soil with large grains

◀ *Sandy soil* is loose and easy to dig. Its grains are larger than those in clay and silt. Sandy soil feels rough and gritty. It feels dry because water passes quickly through it. Sandy soil has very few nutrients. Many plants cannot grow well in this kind of soil. Some plants, like these cactuses, have adaptations for growing in dry places. Sandy soil is found in many places, especially in desert areas and near seacoasts.

Lesson 2 Review

1. What materials make up soil?

2. Why is soil important to living things?

3. Describe three main kinds of soil.

4. **Identify the Main Idea**
   Look at the section titled "Soil and Living Things" on page C42. Identify the stated main idea of this section.

Observing What Is in Soil

Process Skills

- observing
- inferring

Materials

- safety goggles
- potting soil
- soil from schoolyard
- white construction paper
- marker
- hand lens
- plastic spoon
- masking tape
- 2 plastic cups with lids
- water
- clock

Getting Ready

In this activity you can find out about the materials that make up soil by examining soil samples.

Follow This Procedure

1 Make a chart like the one shown. Use your chart to record your observations.

Observations of potting soil	Observations of schoolyard soil

2 Put on your safety goggles. Pour a sample of each kind of soil in a separate pile on the white paper. Write the kind of soil near each pile.

3 **Observe** each soil sample. Describe its color and how it feels. Use the hand lens to look at tiny particles and different materials in the soil. Use the spoon to separate the materials (Photo A). Record your observations.

Photo A

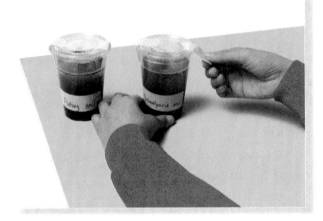

Photo B

④ Use the marker and masking tape to label one cup *Potting soil*. Label the other cup *Schoolyard soil*. Fill the first cup half full with potting soil. Fill the second cup half full with schoolyard soil.

⑤ Fill each cup $\frac{3}{4}$ full with water. Replace the lids. Tape the lids to the cups (Photo B).

 Safety Note *Wipe up any spills immediately.*

⑥ Hold the lids on and shake the cups. Allow the cups to settle for about 15 minutes. Look for large and small soil particles in each sample. Look for humus that may remain floating in the water. Record your observations. Wash your hands after this activity.

Self-Monitoring

Do I have any questions to ask before I continue?

Interpret Your Results

1. Compare and contrast the soil samples.

2. Make an **inference.** Which soil would be better for growing plants? Explain.

Inquire Further

What materials might be found in soils from other areas? Develop a plan to answer this or other questions you may have.

Self-Assessment

- I followed instructions to **observe** soil samples.
- I observed soil samples after they were mixed with water and shaken.
- I recorded my observations.
- I compared and contrasted the soil samples.
- I made an **inference** about which soil would be better for growing plants.

What's the **Big Idea?**

You will learn:

- how to protect resources by recycling and reusing materials.

- how you can reduce the amounts you use.

How Can We Protect Natural Resources?

Don't **DROP** it! You're about to throw an empty juice can in the trash. You can put it someplace else instead. Find out where you can put it so you can save both metal and energy.

Recycling and Reusing Materials

Soil, water, and air are the most important resources. Everyone needs to protect these natural resources because all organisms depend on them. We need to keep them clean. We also need to use them wisely and not waste them.

▲ **New Paper from Newspapers**

Put old newspapers into a recycling bin. They will be used to make new paper. Fewer trees will be cut down if some new paper can be made from old paper.

▲ **Cans and Energy**

Cans are made of aluminum and other metals. Making aluminum from cans uses less energy than making it from ore. So recycling saves ore and fuel!

▲ **Plastic Bags and Bottles**

Some kinds of plastics can be broken up or melted. Then the recycled plastic can be made into new things.

Other resources, such as fuels and ores, are very important too. The earth has only certain amounts of these resources. Once these materials get used up, they will be gone. By recycling, we can help make sure that these materials will last a long time. When something is **recycled,** it's changed so it can be used again.

Each day you might throw away papers, food, cans, and plastic objects. If you put these objects into the trash, someone takes the trash away. Where does the trash go?

In many cities trash gets piled up to form a hill. Machines cover the trash with soil. A place where trash is buried in this way is a **landfill.** However, there are problems with landfills. Some cities are running out of space for them. What can people do to make sure that our soil resources last? You were right if you said "recycle." The pictures on these two pages show some things people can do to protect soil and other resources.

Glossary

recycle (rē sī′kəl), to change something so it can be used again

landfill, a place where garbage is buried in soil

▲ **Reusing**
Many things can be used again. For example, you might use plastic butter tubs to store buttons or to grow small plants. You also can pass clothes you've outgrown to others who can use them.

Making Humus Instead of Trash
Some people actually make humus to add to soil. They do not throw leaves, parts of fruits and vegetables, and grass cuttings into the trash. Instead, they put them in a special pile outdoors. The mixture is called compost. After a time, it decays and becomes humus. Other people put these materials into special bags. The bags then go to a city compost center instead of a landfill. ▶

Reduce Amounts You Use

Another good way to protect resources is to use less of them. If you reduce the amounts of resources you use, they will last longer. If you use less material, you also will have less trash to throw away.

Clean water is a resource that we need to protect. We also have to **conserve** it. That is, we have to keep it from being used up. The chart shows some things we all can do to conserve this resource.

Ways to Conserve Water

- Turn off the water when you brush your teeth.

- Take short showers instead of baths.

- If you take baths, fill the tub only half full.

- Check all the faucets in your home for leaks. Tell an adult about any leaks you find.

- Sweep sidewalks with a broom instead of using water and a hose.

- If you wash a bicycle or a car, turn off the hose except while you're rinsing. Wash a bicycle on the grass. Then you give water to the grass instead of the sidewalk.

Lesson 3 Review

1. How can recycling help protect soil?

2. What are some ways to conserve water?

3. **Multiply Tens**
   Recycling one aluminum can saves enough energy to light an electric light bulb for 210 minutes. Recycling 10 cans will save enough energy to light the bulb for how many minutes?

Experimenting with Soils and Water

Materials

- safety goggles
- masking tape
- pencil
- 3 plastic foam cups
- 3 pieces of cheesecloth
- large spoon
- sandy soil
- clay soil
- potting soil
- water
- graduated cylinder
- clear plastic cup
- clock with a second hand
- foil pan

Process Skills

- formulating questions and hypotheses
- identifying and controlling variables
- experimenting
- estimating and measuring
- collecting and interpreting data
- communicating

State the Problem

How much water flows through different types of soil?

Formulate Your Hypothesis

Will more water flow through sandy soil, clay soil, or potting soil in one minute? Write your **hypothesis.**

Identify and Control the Variables

The type of soil is the **variable** you can change. You will perform three trials. Use sandy soil in Trial 1, clay soil in Trial 2, and potting soil in Trial 3. Use the same amount of water and soil for each trial. Allow the soils to drain for the same amount of time (one minute).

Continued ➔

Photo A

Photo B

Test Your Hypothesis

Follow these steps to perform an **experiment**.

1 Make a chart like the one on the next page. Use your chart to record your data.

2 Put on your safety goggles. Make three labels on pieces of masking tape. Label one *Sandy soil,* label the second *Clay soil,* and label the third *Potting soil.* Put each label on a plastic foam cup.

3 Use the pencil to punch three holes in the bottom of each cup. Place a piece of cheesecloth in each cup to cover the holes.

4 Use the spoon to fill each cup $\frac{2}{3}$ full with the soil type named on the label (Photo A).

5 Put 50 mL of water into the graduated cylinder. Hold the cup with sandy soil over a clear plastic cup. Pour the water on the soil (Photo B).

 Safety Note *Wipe up any spills immediately.*

6 Have a partner keep time. Let the water go through the soil for one minute. Then put the cup into the foil pan.

7 Pour the water from the clear plastic cup into the graduated cylinder (Photo C). **Measure** how much water dripped through the sandy soil in one minute. **Record** your **data** in your chart.

8 Repeat steps 5–7 with the clay soil and the potting soil. Record your data in your chart.

Photo C

Collect Your Data

Type of soil	Amount of water that dripped through soil
Sandy soil	
Clay soil	
Potting soil	

Interpret Your Data

1. Label a piece of grid paper as shown. Use the data from your chart to make a bar graph.

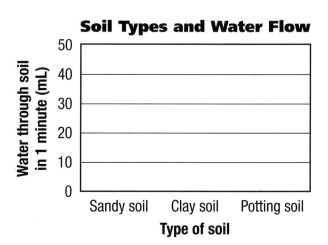

Soil Types and Water Flow

2. Study your graph. How much water flowed through each soil type in one minute?

State Your Conclusion

How do your results compare with your hypothesis? Explain how the type of soil affects how much water passes through it. **Communicate** your results. Compare and contrast your results with the class.

Inquire Further

Which type of soil would be best for growing a tomato plant? Develop a plan to answer this or other questions you may have.

Self-Assessment

- I made a **hypothesis** about how much water flows through different soil types.
- I **identified** and **controlled variables.**
- I followed instructions to perform an **experiment** to test three types of soil.
- I **collected** and **interpreted data** by recording **measurements** and making a graph.
- I **communicated** by stating my conclusion about how the type of soil affects how much water flows through it.

Chapter 2 Review

Chapter Main Ideas

Lesson 1
• Rocks are made of minerals.
• Fossils form from animals and plants that lived a very long time ago.
• Some fuels, called fossil fuels, were formed from the bodies of plants and animals that lived millions of years ago.
• Minerals are kinds of natural resources that can be used for making many things.

Lesson 2
• Soil is a natural resource that is made of weathered rock, decayed organisms, water, and air.
• Soil provides minerals and other nutrients to plants and provides a home for some animals.
• Different kinds of soil have pieces of different size, contain different minerals, and can hold different amounts of water.

Lesson 3
• By recycling and reusing materials, we can help make sure that the materials will last for a long time.
• By conserving resources, such as clean water, we can help make sure that there will be enough for others.

Reviewing Science Words and Concepts

Write the letter of the word or phrase that best completes each sentence.

a. clay soil
b. conserve
c. decay
d. fossil
e. fuel
f. humus
g. landfill
h. loam
i. mineral
j. natural resource
k. nutrient
l. ore
m. recycle
n. sandy soil

1. A material that forms in the earth from nonliving matter is a ____.
2. The hardened parts or marks left by an organism that lived very long ago is a ____.
3. A material that can be burned to produce useful heat is a ____.
4. A material from the earth that living things can use is a ____.
5. A rock that has a large amount of useful materials is an ____.
6. To slowly break down or rot is to ____.
7. The material made of decayed organisms in soil is ____.

8. A material that plants and animals need to live and grow is a ___.

9. A kind of soil that has tiny grains that are packed closely together is ___.

10. A kind of soil that is loose and has large grains is ___.

11. A kind of soil that is good for planting is ___.

12. To change something so it can be used again is to ___.

13. A place where garbage is buried under soil is a ___.

14. To use a resource carefully so it won't be used up is to ___.

Explaining Science

Draw and label a diagram or write a paragraph to answer these questions.

1. What are three main ways that rocks form?

2. How do soils form?

3. What are some things we can do to protect natural resources?

Using Skills

1. Suppose that a school produces about 10 kilograms of garbage a day. **Multiply by tens** to figure out how much garbage this school would produce in a week.

2. Many cans, plastic bottles, and paper products can be recycled. **Observe** where people usually put these materials. Make a poster to remind people to use recycling bins.

3. Organisms take things from soil and they also add things to soil. Make a list to **communicate** to the class what you learned about materials that organisms add to soil.

Critical Thinking

1. Suppose you live near a field where many different plants grow. A sandy beach is also nearby. After a heavy rain, you want to play outdoors in a dry place. Which place would you **infer** to be more dry? Explain your answer.

2. Imagine that you are going to plant some seeds. Why do you **decide** to add humus to the soil?

3. **Compare and contrast** clay soil, sandy soil, and loam.

Space at Your Fingertips!

Imagine getting a close-up view of the sun, moon, and planets. All you need is a computer and the correct Internet address. Soon you're viewing objects in space without even leaving home!

Chapter 3

The Sun, Planets, and Moon

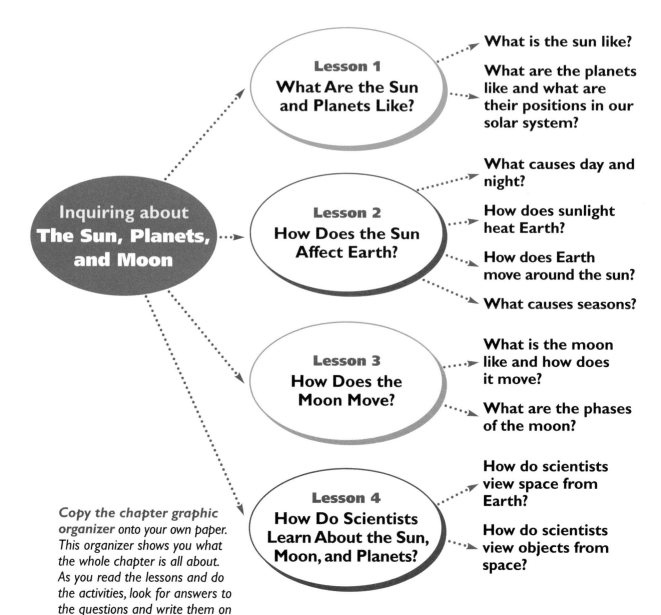

Inquiring about The Sun, Planets, and Moon

Lesson 1
What Are the Sun and Planets Like?

What is the sun like?

What are the planets like and what are their positions in our solar system?

Lesson 2
How Does the Sun Affect Earth?

What causes day and night?

How does sunlight heat Earth?

How does Earth move around the sun?

What causes seasons?

Lesson 3
How Does the Moon Move?

What is the moon like and how does it move?

What are the phases of the moon?

Lesson 4
How Do Scientists Learn About the Sun, Moon, and Planets?

How do scientists view space from Earth?

How do scientists view objects from space?

Copy the chapter graphic organizer onto your own paper. This organizer shows you what the whole chapter is all about. As you read the lessons and do the activities, look for answers to the questions and write them on your organizer.

Exploring Size and Distance of Earth and Sun

Process Skills

- estimating and measuring
- communicating
- making and using models

Materials

- metric ruler
- peppercorn
- posterboard circle
- string

Explore

1 **Measure** and record the diameter of the peppercorn "Earth" and the circle "sun."

2 Take the circle, string, and peppercorn to an area with lots of room. Have a partner hold up the circle sun.

3 Stretch the string out straight from the person holding the sun. Carry the peppercorn Earth to the other end.

4 Stretch the string out from Earth's position. Carry Earth to the other end.

5 Repeat step 4 three more times. This is a **model** of the sizes of Earth and the sun and the distance between them.

Reflect

Communicate. Summarize what you learned about the size of Earth and the sun and the distance from Earth to the sun.

? Inquire Further

How could you model the size and distance from Earth to the moon? Develop a plan to answer this or other questions you may have.

Identifying Supporting Facts and Details

You have already learned that the main idea is the topic of a paragraph or lesson. Supporting facts and details give you more information about the main idea. As you read Lesson 1, *What Are the Sun and Planets Like?*, think of the main idea for the section on the sun. Look for supporting facts and details that tell you more about the sun. Do the same thing for the section on the planets.

Example

One way to remember facts and details is to make a list like the one shown here. Copy the list on a piece of paper. Write three supporting facts and details under each main idea in the list.

A. The sun is an interesting object in the sky.

   1.

   2.

   3.

B. The planets in our solar system are different from each other.

   1.

   2.

   3.

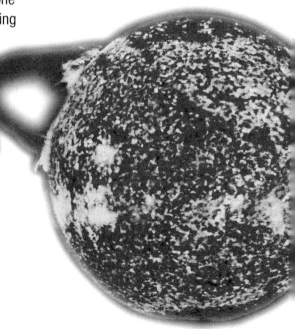

▲ *Have you ever wondered what the sun is like?*

Talk About It!

1. How would supporting facts and details help you better understand something you had just read?

2. What facts and details did you learn when you did the activity *Exploring Size and Distance of Earth and Sun?*

What Are the Sun and Planets Like?

You will learn:

- what the sun is like.
- what the planets are like and the position of each planet in our solar system.

Oh, good! It's a bright, sunny day. Maybe at sundown you'll see a beautiful sunset. Stop and think. It seems as if the sun plays a big part in our lives. What is this thing called the sun, anyway?

The Sun

Glossary

star, a very large mass of hot, glowing gases

Think about the many stars you might see twinkling in the night sky. A **star** is a ball of hot, glowing gases. Look at the picture of the sky during the day. You see only the sun. Where did all of the stars go? Because the sun is so bright, you cannot see other stars during the day.

Like the dots of light you see at night, the sun is also a star. It seems larger and brighter than other stars because it is the closest star to Earth. Other stars look so small because they are very far away from Earth.

◄ *The sun is the closest star to Earth.*

C60

Notice in the picture below how the surface of the sun seems to glow. The temperature on the surface of the sun is very hot, about 5,500°C. The temperature in the center of the sun is much hotter. The sun produces energy in its center. This energy travels out to the surface of the sun. Energy from the sun lights and warms Earth.

There are dark spots and bright spots on the sun's surface. The dark spots are called sunspots. Sunspots appear dark because they are cooler than the areas around them. The temperature in the bright spots, called flares, is higher than the temperature of the sun's surface. The loop on the side of the sun is a huge, bright arc of gas that rises from the edge of the sun, then flows back into it.

▼ *The surface of the sun*

Glossary

planet (plan′it), a large body of matter that moves around a star such as the sun

solar system (sō′lər sis′təm), the sun, the planets and their moons, and other objects that move around the sun

crater (krā′tər), a large hole in the ground that is shaped like a bowl

The Planets

A **planet** is a large body of matter that moves around a star such as the sun. Read about the nine known planets on these two pages.

The sun, the nine planets and their moons, and other objects in the sky make up our **solar system**. Everything in our solar system moves around the sun.

Jupiter

Jupiter, the fifth planet from the sun, is the biggest planet in our solar system. Jupiter is a large ball of gas. It has at least 16 moons and is surrounded by one thin ring.

Mars

The fourth planet from the sun, Mars, is often called the red planet because its surface is a reddish color. Mars is a rocky desert with volcanoes, valleys, and large dust storms. It has two moons.

Mercury

*The planet closest to the sun is called Mercury. Mercury is hot, dry, and covered with craters. A **crater** is a large hole in the ground shaped like a bowl. Mercury has no moons.*

Earth

From space, Earth looks like a white and blue ball. Our planet is the third planet from the sun. It has the air and water needed to support life. Earth has one moon.

Venus

Venus, the second planet from the sun, is surrounded by a thick layer of clouds. It is a hot, rocky planet covered with mountains, valleys, and craters. Venus has no moons.

Pluto

Pluto is usually the ninth and farthest planet from the sun. It is the smallest planet in our solar system. Pluto is a hard, frozen planet. It has one moon.

Saturn

Saturn, like Jupiter, is a giant ball of colored gas. It has at least 18 moons and many rings. Saturn is the sixth planet from the sun.

Neptune

The eighth planet, Neptune, is a large, cold planet made of gases. It is a pale blue color and has rings that are very hard to see. Neptune has eight moons.

Uranus

Uranus is also a ball of gas. It is the seventh planet from the sun. From space, Uranus looks blue-green in color. It has a group of rings that are very close together. Uranus is different from the other planets because it lies on its side. It has at least 15 moons.

Lesson 1 Review

1. What are three things you know about the sun?

2. What are the names of the planets in our solar system, from closest to the sun to farthest away?

3. Facts and Details
Which four planets in our solar system are made of gas?

Observing Shadows Caused by Sunlight

Process Skills

- collecting and interpreting data
- observing
- estimating and measuring

Materials

- piece of cardboard
- black marker
- clay
- straw (one half)
- clock
- metric ruler

Getting Ready

In this activity you can find out how shadows caused by sunlight change during the day.

Follow This Procedure

1 Make a chart like the one shown. Use your chart to record the time and length of the shadows.

2 Write *North* at the top of the cardboard.

Time	Length of shadow

3 Place the piece of cardboard in a sunny spot on the ground. Make sure the top of the cardboard is pointing north.

 Safety Note *Do not look directly at the sun.*

4 Press the clay lightly onto the bottom center of the cardboard. Place the straw piece in the middle of the clay. Make sure the straw is standing straight up and down in the clay (Photo A).

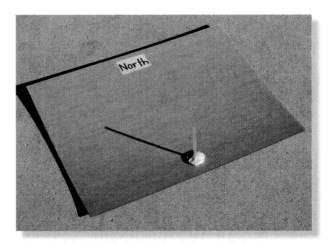

Photo A

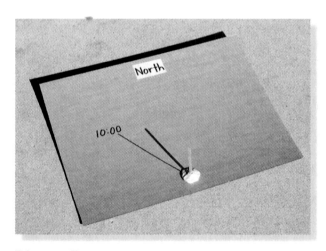

Photo B

⑤ **Collect data** by recording the time on your chart.

⑥ **Observe** the shadow on the cardboard. Use your marker to draw a line on the cardboard from one end of the shadow to the other end. Write the time at the end of the shadow (Photo B).

Self-Monitoring
Have I drawn the shadow from the base of the clay to the tip of the shadow?

⑦ **Measure** the length of the shadow. Record your measurement.

⑧ Repeat steps 5–7 every hour for several hours. Be sure your cardboard is in the same position each time.

Interpret Your Results

1. Describe how the position of the shadows changed.

2. What can you conclude about the time of day and the length and position of the shadows?

⑦ Inquire Further

How could you make the same shadows indoors using the cardboard, clay, straw, and a flashlight? Develop a plan to answer this or other questions you may have.

Self-Assessment

- I followed instructions to **observe** shadows from the sun.
- I **collected data** by recording the time of each shadow.
- I **measured** the length of each shadow.
- I recorded my observations and measurements.
- I **interpreted data** from the chart and made a conclusion about the time of day and the length and position of the shadows.

You will learn:
- **what causes day and night.**
- **how sunlight heats Earth.**
- **how Earth moves around the sun.**
- **what causes seasons.**

Glossary

axis (ak′sis), an imaginary straight line through the center of Earth around which Earth rotates

rotate (rō′tāt), to spin on an axis

Lesson 2

How Does the Sun Affect Earth?

You put your wet, muddy sneakers outside in the sun to dry. An hour later, they are in the shade. What's going on here?

Day and Night

Notice the line drawn through the center of Earth. This imaginary line is called an **axis.** Earth spins, or **rotates,** on its axis. You cannot feel Earth rotating, but it makes one full rotation each day, or every twenty-four hours.

Earth's rotation causes day and night. It is daytime on the side of Earth that faces towards the sun. The side of Earth that faces away from the sun has nighttime. An area on the surface of Earth rotates into sunlight, then into darkness, and then back into sunlight.

During the day, the sun seems to move across the sky. However, the sun does not really move at all. Instead, Earth is moving. Earth is rotating on its axis, which only makes it seem as if the sun is moving. ▶

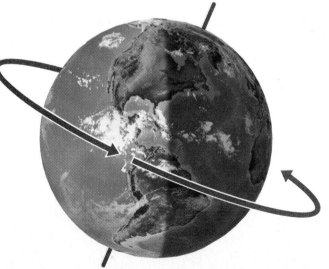

Heating Earth

Almost all life on Earth depends on light from the sun. Plants could not make sugars without sunlight. Animals would have no food or oxygen. Life on Earth would not survive.

Light from the sun warms Earth. Look at the picture below. Notice that sunlight hits some parts of Earth straight on. It hits other parts of Earth at an angle. Places where sunlight hits Earth straight on are warmer than places where it hits Earth at an angle. The picture shows that the middle part of Earth is warmer than other areas on Earth.

Light from the sun hitting Earth ▼

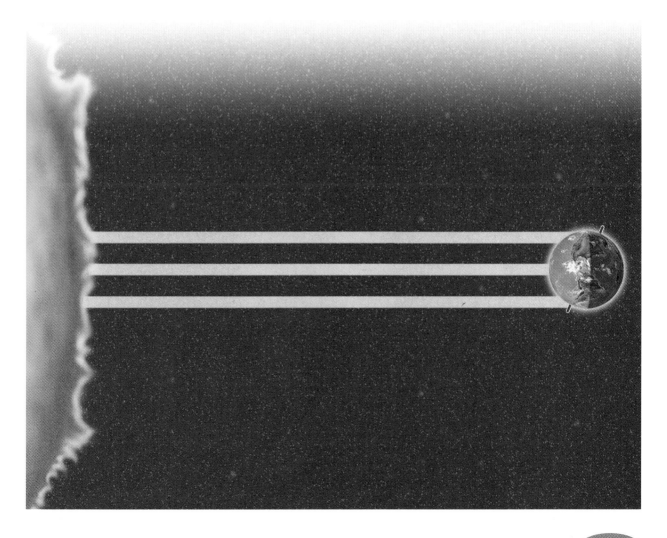

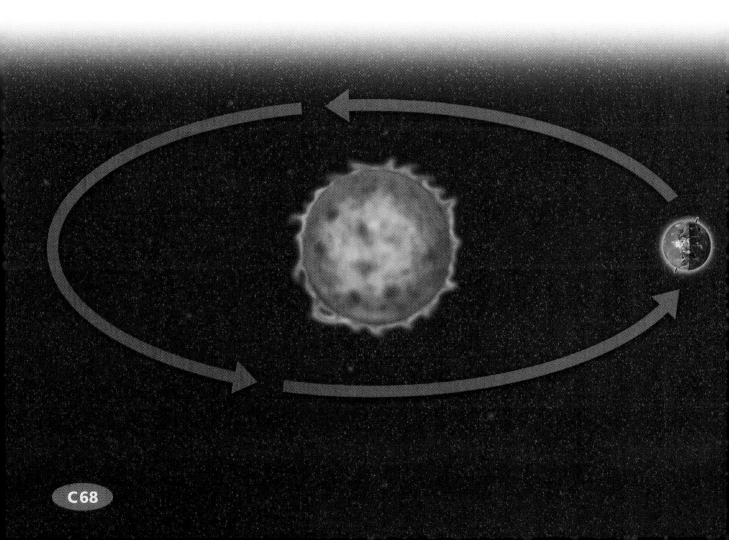

Glossary

orbit (ôr′bit), the path an object follows as it moves around another object

revolution (rev′ə lü′shən), movement of an object in an orbit around another object

Earth's Orbit

Trace the orbit of Earth around the sun with your finger. Notice that Earth's orbit around the sun is almost a circle. ▼

How Earth Moves Around the Sun

At the same time Earth is rotating on its axis, it is also traveling around the sun, as shown below. The path Earth travels around the sun is called an **orbit.** One complete orbit around the sun is called a **revolution.** It takes one year, or about 365 days, for Earth to make one revolution.

Earth is constantly rotating and revolving. Remember that the rotation of Earth causes day and night. One complete rotation takes one day, or twenty-four hours. One complete revolution of Earth around the sun takes one year, or about 365 days.

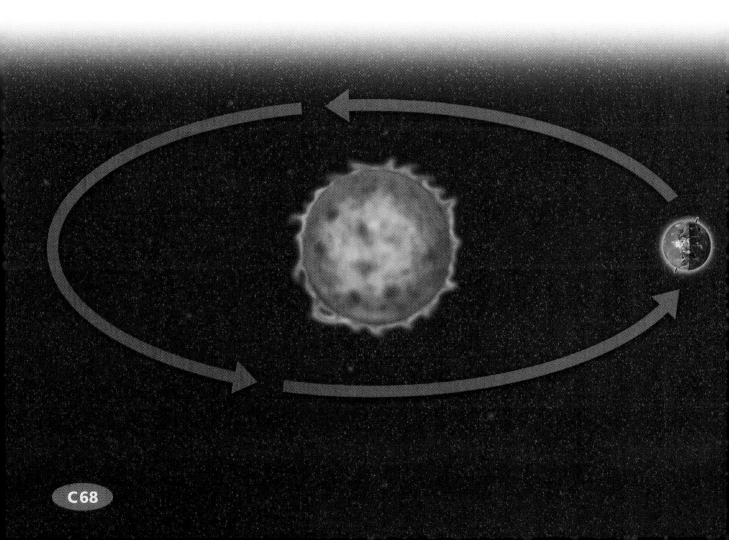

How do people keep track of time during a year? They look at a calendar, of course! Look at the calendars on this page. Throughout the years, people often used the movement of the sun, moon, and stars to track time. Many times, this information was organized into a calendar. Calendars help people know important times of the year, such as when to plant crops or when to prepare for cold weather.

A Modern Calendar

Today we use a calendar that has 12 months in a year. Each month has either 30 or 31 days, except February, which usually has 28 days. Every fourth year, we have a leap year of 366 days. We add the extra day to February. In a leap year, February has 29 days. ▼

▲ An Ancient Aztec Calendar

The Aztecs lived in Mexico hundreds of years ago. One part of the Aztec calendar had 365 days in a year. The year consisted of 18 months. Each month had 20 days. The Aztecs added 5 extra days during the year to complete their calendar. This calendar worked for only 52 years. Then the calendar no longer matched the seasons. When this happened, the Aztecs would start another 52-year calendar.

Seasons

Look at the picture. Notice that Earth's axis is not straight up and down. It is slightly tilted. As Earth revolves around the sun, its axis is always tilted in the same direction. During the year, the part of Earth tilted toward the sun gets more direct sunlight. The part of Earth tilted away from the sun gets less direct sunlight.

March

In March, the north and south halves of Earth are getting about the same amount of sunlight. The north half of Earth is getting warmer. The south half is getting cooler.

June

The north half of Earth is tilted towards the sun. The south half is tilted away from the sun. The north half of Earth gets more direct sunlight than the south half. It is summer on the north half of Earth and winter on the south half.

September

As in March, the north half and south half of Earth get about the same amount of sunlight. The north half of Earth is getting cooler, while the south half is getting warmer.

It is the tilt of Earth's axis and Earth's revolution around the sun that cause the seasons. A season lasts about three months. Four seasons—spring, summer, autumn, and winter—make one year.

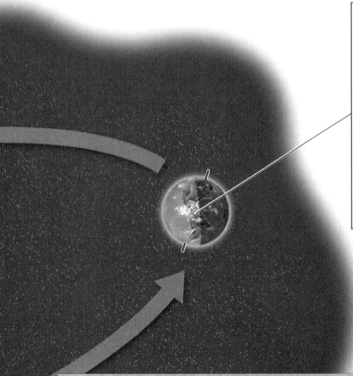

December
The north half of Earth is tilted away from the sun. Sunlight is hitting the north half of Earth at more of an angle. It is winter here. The south half of Earth is getting more direct sunlight because it is tilted towards the sun. It is summer here.

Lesson 2 Review

1. What causes day and night?

2. How does the way sunlight hits Earth affect the way different parts of Earth are heated?

3. How does Earth move during one year?

4. What causes seasons?

5. **Draw Conclusions**
   The north half of Earth is tilted away from the sun. What season is it on the north half of Earth? Why? What season is it on the south half of Earth? Why?

Modeling the Moon's Motion and Light

Process Skills

- making and using models
- observing
- inferring

Materials

- masking tape
- plastic-foam ball
- lamp

Getting Ready

You can find out how the moon moves and how it appears to change shape in the sky by making a model.

Darken the room to see the light from the lamp clearly.

Follow this Procedure

1 Make a chart like the one shown. Use your chart to record your drawings of the moon.

	Drawings of moon
Facing to the left	
Facing the light	
Facing to the right	
Facing away from the light	

2 Use the masking tape to make an X on the ball. The ball is a **model** of the moon.

3 The lamp represents the sun and you represent Earth. Face away from the lamp. Hold up the ball so it is lit by the lamp. The X should be facing you (Photo A).

⚠ **Safety Note** *Do not look directly into the light.*

4 Slowly make a quarter turn to the left. Keep the X facing you. **Observe** what happens to the light on the ball. A shadow should appear on one side of the ball (Photo B). Draw the ball and shade in the part that is in the shadow.

Photo A

Photo B

5 Repeat step 4. You should now be facing the lamp. How much of the ball appears to be in the shadow?

6 Repeat step 4 two more times, until you are standing in the position where you started.

Self-Monitoring
Did I keep the X in the same position? Did I make drawings of the ball in all four positions?

Interpret Your Results

1. What happened to the shadow on the ball as you turned to face the lamp? What happened to the shadow as you turned away from the lamp?

2. The X on the ball was always facing you. Did you ever see the other side of the ball as you turned? Make an **inference.** Do people ever see the other side of the moon from Earth? Explain.

**Inquire Further**

How can you model a lunar eclipse? Develop a plan to answer this or other questions you may have.

Self-Assessment

- I followed instructions to **model** how the moon appears to change shape.
- I **observed** the shadow on the ball in different positions.
- I drew pictures of the ball in each position.
- I described the movement of the moon and its shadow.
- I made an **inference** about the way the moon appears to people on Earth.

You will learn:

- what the moon looks like and how it moves.
- what the phases of the moon are.

Lesson 3

How Does the Moon Move?

Some people joke that the moon is made of green cheese. Or some people say they see a man in the moon. Over the years, people have imagined many things about the moon.

Ways the Moon Moves

Look at the picture of the moon. Earth is about four times the size of the moon. Like Earth, the surface of the moon has mountains and valleys. The moon is also full of craters and large, flat plains covered with rocks and dust.

The moon has no air. Scientists used to think that the moon had no water either. Recently, however, ice made from water was discovered on the moon. Scientists do not know where this ice came from. One thought is that objects that hit the moon contained ice. Most of the ice quickly changed into water vapor. However, some ice landed in craters that were always in the shade. This ice did not change into water vapor. It remained ice.

The moon in a night sky ▼

The moon moves in much the same way as Earth. Like Earth, the moon rotates on its axis. The moon is a satellite of Earth. A **satellite** is an object that revolves around another object. Notice in the picture how the moon revolves around Earth in much the same way Earth revolves around the sun. It takes about one month for the moon to revolve once around Earth.

The moon reflects light from the sun. Sunlight is reflected off the part of the moon that faces the sun. The part of the moon that does not face the sun is dark. You see only the lighted part of the moon that faces Earth. You see different amounts of the moon's lighted part as it revolves around Earth.

The pull of the moon's gravity is the main cause of tides on Earth. A **tide** is the rise and fall of water along an ocean shore. Notice the two kinds of tides in the pictures.

Glossary

satellite (sat′l īt), an object that revolves around another object

tide, the rise and fall of the ocean mainly due to the moon's gravity

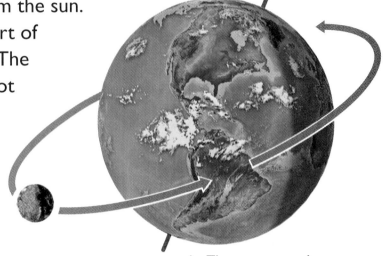
▲ The moon revolves around Earth.

▲ High tide

▲ Low tide

The Phases of the Moon

Glossary

phase (fāz), the shape of the lighted part of the moon

Night after night, the moon seems to change shape in the sky. Sometimes it looks like a bright circle. Other times you can see only a small part of the moon. Still other times you do not see the moon at all.

The moon does not change shape. It only appears to change shape because all you see is the lighted part. All these different shapes together are called the **phases** of the moon. You see the moon's phases shown in the pictures on the next page as it revolves once around Earth.

Charting the Phases of the Moon

How do people know when each phase of the moon appears in the sky? They look up! Many calendars show the phases of the moon. You can make your own calendar showing the phases of the moon by observing the night sky every day for a month. ▶

▲ New Moon

Sometimes, as the moon revolves, it gets between the sun and Earth. Then the dark half of the moon faces towards Earth. The lighted part of the moon faces away from Earth. You cannot see any moon in the sky. This phase is called a new moon.

▲ Crescent Moon

You can see more and more of the moon's lighted part as it revolves. A night or two after the new moon, you see only a small piece of the lighted part of the moon. This phase is called a crescent moon. You usually see a crescent moon shortly after the sun sets.

▲ Half Moon

It is now about a week after the new moon. The moon looks like a half circle. This phase is sometimes called a half moon.

▲ Full Moon

About a week after the half moon, you can see all of the lighted part. The moon looks like a complete circle. This phase is called a full moon. After the full moon, you begin to see less of the moon each night. About two weeks later, there is another new moon.

Lesson 3 Review

1. How does the moon move?

2. List four phases of the moon.

3. Facts and Details
List three facts you know about the moon.

You will learn:

- how scientists view space from Earth.
- how scientists view objects from space.

telescope (tel′ə skōp), an instrument for making distant objects appear nearer and larger

Lesson 4

How Do Scientists Learn About the Sun, Moon, and Planets?

I want to see more! You think about the moon and all the stars and planets out there in the sky. All you can see is the moon and tiny bits of twinkling lights far, far away. You want a closer look at what's out there!

A View from Earth

The child in the picture is using a telescope to look at the sky. A **telescope** is an instrument used for making objects appear nearer and larger.

People have always been interested in objects in the sky. For a long time, the only way people on Earth could observe the sky was with their eyes. They thought Earth was the center of the solar system and everything in the sky revolved around Earth.

◀ Using a telescope

Over four hundred years ago, a man named Galileo used a telescope to look more closely at the sky. Galileo was the first person to see mountains and valleys on the moon. He saw thousands more stars in the night sky than you can see with just your eyes. Galileo also saw four moons orbiting Jupiter. He observed that Venus has phases like our moon. The picture shows Galileo looking at the sky through a telescope. The telescope helped Galileo and other scientists prove that the sun, not Earth, is the center of the solar system.

The building in the picture below is called an observatory. Scientists use the large, powerful telescopes in an observatory to study space.

▲ *Galileo used a telescope to observe the sky.*

◀ *Keck Observatory on Mauna Kea, Hawaii*

C79

Glossary

astronaut
(as'trə nȯt), a person
who travels in space

A View from Space

You have learned that people use telescopes to study space from Earth. Scientists use spacecraft to study the moon and planets from space. Cameras, computers, and other equipment on spacecraft record information that scientists cannot gather from Earth.

Some spacecraft carry astronauts. An **astronaut** is a person who travels in space. Astronauts and spacecraft send information back to Earth. Read about information gathered by astronauts and spacecraft on these two pages. Scientists study this information to learn more about the sun, moon, and planets.

▲ The Moon

Astronauts have traveled to the moon and walked on it six times. Each time, they explored a different part of the moon. They saw craters and brought back moon rocks for scientists to study.

Hubble Space Telescope

Hubble Space Telescope is a large observatory in the sky. Its powerful telescope can see objects and events that are invisible to the strongest telescopes on Earth. Hubble continuously orbits Earth, sending back new and important information about objects and events in space. ▶

◀ Uranus
The Voyager 2 spacecraft took pictures of Uranus and its rings and moons.

▲ Venus
Pictures taken by spacecraft showed Venus's rocky surface. They also showed that Venus has no oceans, lakes, or rivers.

▲ Mars

Spacecraft without astronauts have landed on Mars. Cameras from the spacecraft took pictures that showed Mars has mountains and valleys. Scientists recently sent the spacecraft Pathfinder to Mars. They want to learn more about the rocks and soil on the planet.

▲ Cassini Probe

The Cassini Probe is on a mission to explore Saturn and its rings and satellites.

Saturn ▶

Spacecraft have orbited Saturn. Scientists learned more about Saturn's rings and moons from pictures taken by Voyager 2. They hope to learn even more from the Cassini Probe.

Lesson 4 Review

1. How do telescopes help scientists study the sun, moon, and planets?

2. How do spacecraft help scientists learn about the planets?

3. **Facts and Details**
   What are three things scientists have learned about the planets from information gathered by spacecraft?

▲ Jupiter

Without pictures taken by cameras on the spacecraft Voyager, scientists would not have known that Jupiter was surrounded by one very thin ring.

Chapter 3 Review

Chapter Main Ideas

Lesson 1
• The sun is a ball of hot, glowing gases and Earth's most important source of energy.
• The nine planets in our solar system are Mercury, Venus, Earth, Mars, Jupiter, Saturn, Uranus, Neptune, and Pluto.

Lesson 2
• Earth's rotation on its axis causes day and night.
• Places where sunlight hits Earth straight on are warmer than places where it hits Earth at an angle.
• Earth revolves in an orbit around the sun.
• As Earth revolves, the tilt of Earth's axis causes seasons.

Lesson 3
• The moon rotates on its axis and revolves around Earth.
• The phases of the moon can be seen as the moon makes one revolution around Earth.

Lesson 4
• Telescopes have helped scientists study objects in the sky from Earth.
• Spacecraft have helped scientists learn more about objects in space.

Reviewing Science Words and Concepts

Write the letter of the word or phrase that best completes each sentence.

a. astronaut
b. axis
c. crater
d. orbit
e. phase
f. planet
g. revolution
h. rotate
i. satellite
j. solar system
k. star
l. telescope
m. tide

1. The sun is the closest ___ to Earth.
2. A body of matter that moves around the sun is called a ___.
3. Our ___ consists of the sun, the nine planets and their moons, and other objects in the sky.
4. A large hole in the ground that is shaped like a bowl is called a ___.
5. Earth spins on an imaginary line called an ___.
6. It takes Earth twenty-four hours to ___ on its axis.
7. Each complete orbit Earth makes around the sun is called a ___.

8. The moon is called a ____ of Earth because it revolves around Earth.

9. A ____ is the rise and fall of water along an ocean shore.

10. A crescent moon is one ____ of the moon.

11. You can use a ____ to make objects appear nearer and larger.

12. An ____ is a person who travels in space.

13. Earth travels in a path called an ____ around the sun.

Explaining Science

Draw and label a diagram or write a short answer to explain these questions:

1. What are the names of the nine planets in our solar system?

2. What causes night and day?

3. What are the different phases of the moon? What does the moon look like in the sky during each phase?

4. What do scientists use to help them study space from Earth?

Using Skills

1. Write a paragraph giving **supporting facts and details** about the following main idea: Telescopes have helped people learn about the sky.

2. Venus does not take as long to revolve around the sun as Saturn does. Write down what you **infer** is the reason for this difference.

3. You **observe** a full moon shining brightly in the night sky. Write a short statement explaining why the moon shines.

Critical Thinking

1. Compare and contrast Earth and Uranus. How are they alike? How are they different?

2. Based on what you know about the distance of the planets from the sun, **draw a conclusion** about why Mercury and Venus are hotter than Earth.

3. Suppose Earth made one complete rotation on its axis every eight hours. **Predict** how many hours would pass from the start of one day to the start of the next day.

It's Raining! It's Pouring!

The sky is covered with low, gray clouds. Rain has been falling for days. It seems as if it will never be dry outside again.

Chapter 4
Clouds and Storms

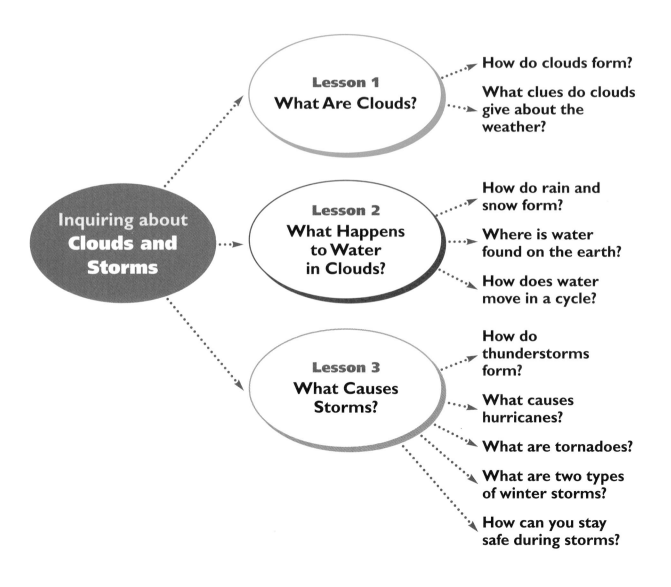

Inquiring about Clouds and Storms

Lesson 1
What Are Clouds?

How do clouds form?

What clues do clouds give about the weather?

Lesson 2
What Happens to Water in Clouds?

How do rain and snow form?

Where is water found on the earth?

How does water move in a cycle?

Lesson 3
What Causes Storms?

How do thunderstorms form?

What causes hurricanes?

What are tornadoes?

What are two types of winter storms?

How can you stay safe during storms?

Copy the chapter graphic organizer onto your own paper. This organizer shows you what the whole chapter is all about. As you read the lessons and do the activities, look for answers to the questions and write them on your organizer.

Exploring How Clouds Form

Process Skills

- making and using models
- observing

Materials

- clear plastic container with lid
- warm water
- 2 ice cubes
- clock with a second hand

Explore

1 To **model** how clouds form, first fill the plastic container about $\frac{2}{3}$ full of warm water.

2 Put the lid on the container. Place 2 ice cubes in the center of the lid as shown. **Observe** any changes inside the container. Record your observations.

3 Observe the container every three minutes for nine minutes. Record your observations.

Reflect

1. Summarize what happened when the lid and ice were placed on the container of warm water.

2. Clouds can form when warm air is cooled. How does this model show the way clouds form?

? Inquire Further

What would happen to the model if you used ice water instead of warm water? Develop a plan to answer this or other questions you may have.

Reading Bar Graphs

How much rain fell each month during the year? You can tell which month had more rain by reading a **bar graph.**

A bar graph is another way to help you compare data. Bar graphs use bars to show data. In this graph, each month has its own bar.

You can find which month had 18 centimeters of rain by looking at the **scale.**

Find the number 18 on the scale. Find the bar that ends at that number. The month with 18 centimeters of rain was May.

Math Vocabulary

bar graph, a graph that uses bars to show data

scale, the numbers that show the units used on a bar graph

Math Tip

Think of each bar as a ruler that measures the amount of rain.

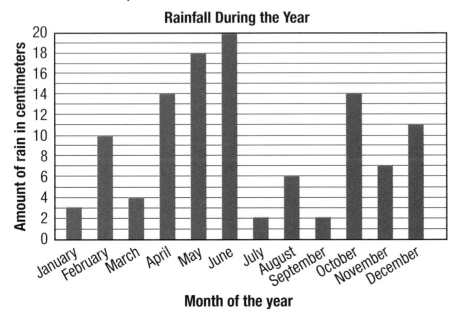

Rainfall During the Year

Talk About It!

1. How can you use the scale to find out which month had the most amount of rain?

2. How does a bar graph help you to compare data?

You will learn:
- how clouds form.
- what weather clues you can get from clouds.

Glossary

water vapor (vā′pər), water that is in the form of a gas

evaporate (i vap′ə rāt′), to change from a liquid to a gas

How can you tell this is a cold day? ▼

Lesson 1

What Are Clouds?

Brrr! It's a very cold day. You bundle up to go outside. The cloudy sky looks like a flat, gray sheet. It's so cold that you can see your breath. It forms a small cloud in front of your mouth.

How Clouds Form

Physical Science

The air contains water vapor. You have learned that **water vapor** is water in the form of a gas, which is invisible. How does water get into the air? The sun's heat causes water to evaporate. Remember that **evaporate** means to change from a liquid to a gas. Wind also helps water evaporate.

Water vapor is in your breath too. Your breath is warm. The picture shows that your breath looks like a small cloud on a cold day because the water vapor in it meets the cold air and changes to a liquid.

When warm air rises or meets cold air, it cools. As the warm air cools, the water vapor **condenses,** or changes from a gas to a liquid. Water vapor condenses onto tiny bits of dust, smoke, and salt in the air to form tiny liquid water droplets. These tiny water droplets float in the air. If the air gets cold enough, the water droplets may freeze into bits of ice.

Sometimes many water droplets or bits of ice form near each other. The picture shows that when this happens, a cloud forms. A **cloud** is a mass of many water droplets or bits of ice. If the air near the ground is cool, low clouds may form in the sky. In warm air, water vapor may rise very high in the sky before it cools, condenses, and forms clouds.

Glossary

condense
(kən dens′), to change from a gas to a liquid

cloud , a mass of many water droplets or bits of ice that float in the air

Glossary

Forming Clouds

Water vapor condenses and clouds form.

Air cools.

Warm air rises.

Weather Clues from Clouds

Clouds can give clues about the weather. They can tell you what kind of weather is coming. Different clouds bring different types of weather. Look at the clouds in the pictures on these two pages. Read about each type of cloud to learn what kind of weather it will bring.

◀ **Towering Storm Clouds**

Huge, tall clouds like these can pile up fast on hot summer afternoons. They can also pile up when cooler air moves toward you. They may have fluffy, white tops, but their bottoms will be dark gray and low. Expect heavy showers, thunderstorms, or even a tornado.

Low, Gray Clouds

These low, gray clouds form in layers. They cover the sky like smooth, even sheets. Sometimes they are so thick they block out much of the sun. Clouds that look like this might bring rain or snow that could last for days. ▶

▲ **Fluffy, White Clouds**

These bright, white heaps of clouds float low in the sky. They have flat bottoms and rounded tops. This type of cloud floats in groups in a blue sky. You see them when the weather is fair and sunny, with no rain.

▲ **High, Feathery Clouds**

These patchy clouds form high in the sky where the air is very cold. Made mostly of tiny bits of ice, they look wispy, like feathers. You see these clouds on clear, sunny, dry days. They can be a sign of warmer air moving toward you.

Lesson 1 Review

1. How does a cloud form?

2. Describe four kinds of clouds and the weather each may bring.

3. Draw Conclusions
You see large, dark gray clouds in the sky. What kind of weather is probably coming your way?

Investigating Clouds and Weather

Process Skills

- making and using models
- observing
- classifying

Materials

- cotton balls
- posterboard or heavy paper
- glue

Getting Ready

In this activity you will be making a poster model of different cloud types. You will also record cloud and weather observations in a chart.

Follow This Procedure

❶ Make a chart like the one shown. Use the chart to record your observations.

❷ Use cotton balls to make a **model** of different cloud types. Review pages C90 and C91 to see how different clouds look. Pull the cotton into different shapes to look like the different cloud types.

Day	Cloud types	Weather observations
1		
2		
3		
4		
5		
6		
7		
8		
9		
10		

❸ Glue some high feathery clouds near the top of a posterboard or sheet of heavy paper.

❹ Glue some fluffy white clouds near the middle of the poster.

❺ Glue some flat cloud layers under the fluffy white clouds.

6 Glue some towering storm clouds at one side of the poster.

7 Write the name of each cloud type next to each model cloud. Also, write the kind of weather you might have when you see that cloud (fair weather, rain or snow, thunderstorms).

8 **Observe** the clouds in the sky and the weather at the same time and place for ten days. **Classify** the clouds into the cloud types on your chart. Record your observations in your chart.

Self-Monitoring
Did I complete each step? Did I observe the sky and weather each day?

Interpret Your Results

1. What cloud types did you see over ten days? What weather came with the types of clouds you saw?

2. Compare and contrast the real clouds with the clouds on your chart.

Inquire Further

Besides watching clouds, what are some other ways to track changes in the weather? Develop a plan to answer this or other questions you may have.

Self-Assessment

- I followed instructions to make a poster **model** of clouds.
- I wrote the types of clouds and the weather that comes with those clouds on my model.
- I **observed** clouds and weather at the same time every day for ten days.
- I recorded my observations and **classifications** of clouds and weather.
- I compared my observations of real clouds with my cloud model.

What's the Big Idea?

You will learn:

- how rain and snow form.
- where water can be found on the earth.
- how water moves in a cycle.

▲ *A rain gauge is a tool that measures rainfall.*

Lesson 2

What Happens to Water in Clouds?

It's really raining hard outside. As a matter of fact, it's pouring! Where did all those water droplets come from?

Rain and Snow

Remember that clouds are made of many tiny water droplets. Often clouds form high in the air where temperatures are below freezing. Then the droplets freeze into tiny bits of ice. When the bits of ice become too heavy to float in the air, they fall through the clouds. If the air temperature near the ground is above freezing, the ice will melt and fall as rain. Notice the rain gauge used to measure rainfall. If the air temperature from the clouds to the ground stays cold enough, the ice will fall as snow.

Water that falls from clouds to the ground is called **precipitation**. Notice that rain and snow are two kinds of precipitation.

◀ *Snow may fall in freezing air.*

Rain usually starts out as snow. ▶

Water on the Earth

Oceans, lakes, rivers, and streams are just a few places where water can be found on the earth. Did you know that oceans contain most of the water on the earth? However, you cannot drink ocean water because it is salt water. Most living things on the earth need fresh water to live.

Lakes, rivers, streams, and ponds contain fresh water. Other fresh water can be found underground. Underground water is called groundwater. People drill wells to bring groundwater to the surface of the earth. Sometimes groundwater flows out of a small opening in the earth's surface. This forms a spring. Spring water may flow into a creek, stream, or river. Is the water in the pictures fresh water or salt water?

▲ *Oceans contain most of the water on the earth. Water covers three times more of the earth's surface than land does.*

Water runs downhill in streams. The streams join larger bodies of water such as rivers, lakes, or the ocean. ▼

Glossary

water cycle
(sī⁄kəl), movement of water from the earth to the air and back to the earth

The Water Cycle

Why does the earth never run out of water? Nature uses water over and over again. This process of using and reusing water is called the water cycle. In the **water cycle ,** water evaporates, condenses, falls as precipitation, and eventually evaporates again. The water cycle repeats over and over. Read about the steps in the water cycle on these two pages.

Condensation
When the air cools, the water vapor condenses into tiny water droplets. The droplets form a cloud.

Evaporation
Energy from the sun evaporates some water from the bodies of water on the earth. Wind also helps evaporate some water. The water enters the air as water vapor.

Precipitation

Precipitation falls from the clouds to the earth's surface. Some water soaks into the ground and becomes groundwater. Water from rivers, streams, and groundwater may reach lakes and oceans. The water evaporates and the water cycle process repeats itself.

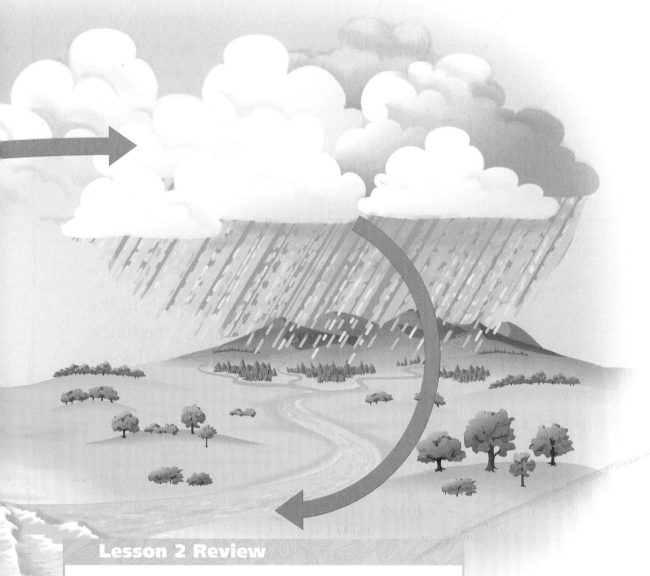

Lesson 2 Review

1. How do rain and snow form?

2. Where are four places on the earth water can be found?

3. What are the steps in the water cycle?

4. **Read Bar Graphs**
   Look at the bar graph on page C87. Which months had the same amount of rain?

Making a Model Tornado

Process Skills

Process Skills

- making and using models
- observing
- communicating

Materials

- tall, clear plastic bottle with cap
- water
- metric ruler
- spoon
- salt
- dishwashing detergent
- food coloring
- glitter or bits of paper

Getting Ready

In this activity you will be making a model to see how air moves in a tornado.

Follow This Procedure

1 Make a chart like the one shown. Use your chart to record your descriptions and drawings.

Speed	Descriptions	Drawings
Rapid swirling		
Slow swirling		
Swirling with glitter or bits of paper		

2 Begin making a **model** tornado by filling the bottle with water until it is about 4 cm from the top.

Safety Note *Wipe up any spills immediately.*

3 Add 1 spoonful of salt to the water. Put the cap on the bottle. Shake the bottle until all the salt dissolves. Add one drop of dishwashing detergent.

4 Firmly screw the cap onto the bottle. Hold the bottle near the bottom and move it rapidly in a swirling motion (Photo A). Immediately **observe** what happens inside the bottle. Write a description and draw a picture of what you observe in your chart.

5 Swirl the model slowly. Describe and draw what you see at each speed.

Photo A

Photo B

6 Remove the cap. Add one drop of food coloring. Add a small pinch of glitter or some bits of paper to the model (Photo B). Replace the cap. Then swirl the bottle rapidly again. Describe and draw what you observe.

Interpret Your Results

1. Communicate. Explain how the liquid inside the bottle moved.

2. What happened when you swirled the bottle quickly? What happened when you swirled the bottle slowly?

3. Describe the motion of the glitter or bits of paper as it moved in the funnel.

Inquire Further

What would happen if you added larger or more massive objects to your model? Develop a plan to answer this or other questions you may have.

Self-Assessment

- I followed instructions to **make a model** of a tornado.
- I made **observations** of my model.
- I recorded descriptions and drawings of my model being swirled at different speeds.
- I **communicated** by explaining how the liquid moved in the bottle.
- I described the motion of the glitter or paper as it moved in the funnel.

What's the Big Idea?

You will learn:

- how thunderstorms form.
- what causes hurricanes.
- about tornadoes.
- about two types of winter storms.
- how to stay safe during storms.

Lesson 3

What Causes Storms?

FLASH! A streak of lightning lights up the sky. **CRASH!** A roll of thunder breaks the stillness. Quick! Get indoors! It's a thunderstorm.

Thunderstorms

A thunderstorm such as the one shown on the next page is the most common type of storm. Notice in the drawing below that storms form when cold air pushes into warm air. The warm air quickly rises up and then rapidly cools. Large clouds with dark bottoms form. Lots of precipitation, usually in the form of rain, will fall in a small area. After a short while, the precipitation will stop. The air will turn cooler, and the day might turn sunny and dry.

When warm air rises very quickly, cooler air rushes in below it. Use your finger to trace the movement of air. ▶

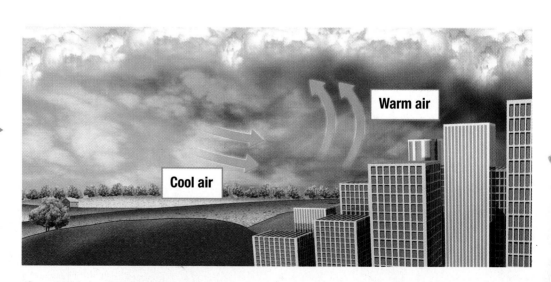

Warm air

Cool air

Many storms, including thunderstorms, are caused by warm air that rises quickly. Think about the last time you saw lightning or heard thunder during a thunderstorm. Lightning flashes when electric charges jump from one cloud to another. You also see lightning when electric charges jump between a cloud and the ground. You hear thunder.

Lightning and thunder signal a storm. Thunderstorms often happen during spring or summer. ▼

Glossary

Glossary

hurricane
(hėr′ə kān), a huge storm that forms over warm ocean water, with strong winds and heavy rains

A **hurricane** is a huge storm that forms over warm ocean water. As air above the ocean warms and rises, huge amounts of water enter the air as water vapor. The water vapor condenses into clouds. Cooler air rushes in to take the place of the warm, rising air. The cooler air warms and rises. As this cycle speeds up, winds blow faster and stronger. Large thunderclouds pile up. Notice in the picture how the clouds begin to spin around and around. Many hurricanes die out over the ocean. The picture of the bending trees shows what can happen when a hurricane makes it to land. A hurricane will die out as it moves over land.

◀ On land, strong hurricane winds can destroy buildings and tear up trees and telephone poles. Heavy rains may also cause floods.

This satellite picture shows a hurricane over the Gulf of Mexico. Notice that the clouds are spinning in a circle. ▶

Tornadoes

A **tornado** is a spinning funnel cloud that reaches from the clouds toward the earth. Often tornadoes like the one shown form after the rain and lightning of a thunderstorm have passed. During a tornado, you would hear lots of noise from the wind.

Tornadoes have the strongest winds on the earth. They move along the earth in a very narrow path. The scientist in the picture is tracking the path of a tornado.

Glossary

tornado (tôr nā′dō), a funnel cloud that has very strong winds and moves along a narrow path

Glossary

▲ A weather scientist tracks a tornado as strong winds push it along its path.

Many tornadoes never touch the ground. If a tornado does touch ground, it can cause much damage. ▼

Glossary

blizzard (bliz′ərd), a snowstorm with strong, cold winds and very low temperatures

Winter Storms

Winter storms can cause freezing temperatures and strong winds. Ice storms often happen when the air temperature is just below freezing, or 0°C. In an ice storm, rain freezes as soon as it hits the ground. An ice storm causes a layer of slick ice to cover the ground, streets, sidewalks, and even trees such as the one in the picture. Walking and driving become difficult and dangerous.

A **blizzard** is a very cold snowstorm with strong winds and heavy snow. In a blizzard, blowing snow can make it difficult to see farther than a short distance. Notice that blowing snow may cause huge snowdrifts.

The weight of ice on tree branches and telephone lines can cause them to break. ▼

Strong, cold winds may pile snow into huge drifts. ▼

Safety During Storms

 Storms can cause damage to many things, including buildings, trees, telephone poles, power lines, and you. Read about some ways you can keep safe during different types of storms.

Thunderstorms

If you are in water, get out.

Do not stand under trees.

Go inside immediately.

Stay away from the television, telephones, water, and metal objects.

Tornadoes

Go to the basement or an inside hall, closet, or bathroom.

Crouch under the stairs or near an inner wall, and cover your head with your arms.

Keep away from windows, water, metal objects, and objects that use electricity.

Winter Storms

Stay indoors if you can.

If you go out, dress warmly.

Wear a hood, mittens, and snow boots.

Hurricanes

Move farther inland.

Go from a mobile home to a public shelter.

Tape windows.

Stay indoors, away from windows.

Lesson 3 Review

1. How does a thunderstorm form?
2. How and where do hurricanes form?
3. What is a tornado?
4. How is an ice storm different from a blizzard?
5. Pick one type of storm and describe how you can stay safe during it.
6. **Cause and Effect**
   What causes most storms?

Chapter 4 Review

Chapter Main Ideas

Lesson 1
• Clouds form when water vapor in the air condenses.
• You can tell what kind of weather is coming by looking at clouds.

Lesson 2
• When the air temperature near the ground is above freezing, bits of ice falling from clouds will melt and fall as rain. If the temperature from the clouds to the ground stays cold enough, the ice will fall as snow.
• Water is found in many places on the earth, including oceans, rivers, lakes, ponds, and underground.
• Water evaporates, condenses, and falls to the earth as precipitation over and over again in the water cycle.

Lesson 3
• A thunderstorm forms when cold air pushes into warm air and the warm air quickly rises.
• A hurricane forms over oceans when large amounts of water enter the air as water vapor, warm air quickly rises, large thunderclouds pile up, and strong, fast winds blow.
• A tornado is a spinning funnel cloud that reaches from the clouds toward the earth.

• Ice storms and blizzards are two types of winter storms.
• People can take steps to keep safe during thunderstorms, hurricanes, tornadoes, and winter storms.

Reviewing Science Words and Concepts

Write the letter of the word or phrase that best completes each sentence.

a. blizzard   **f.** precipitation

b. cloud   **g.** tornado

c. condense   **h.** water cycle

d. evaporate   **i.** water vapor

e. hurricane

1. A funnel cloud that has strong winds and moves along the earth in a very narrow path is called a ___.

2. Water vapor will ___ as warm air cools.

3. The ___ is the movement of water between the ground and the air by evaporation, condensation, and precipitation.

4. A ___ is a storm that forms over warm ocean water and has strong winds and heavy rains.

5. Water in the form of gas is ___.

6. It is very cold with strong winds and lots of snow during a ___.

7. Rain and snow that fall from the clouds to the ground are two kinds of ___.

8. The sun's heat and wind cause water to ___ , or change from a liquid to a gas.

9. When many water droplets or bits of ice form near each other, a ___ will form.

Explaining Science

Draw and label a diagram or write a statement to answer these questions.

1. How do water and air move as clouds are forming?

2. What are the stages of the water cycle?

3. What must happen for a thunderstorm to form?

Using Skills

1. Read the bar graph on page C87 and identify the four months that had the least amount of rain.

2. Suppose a raindrop falls. What **sequence** of steps in the water cycle would the raindrop then follow?

3. Suppose you hear a tornado siren going off in the distance. **Draw a conclusion** about which place would be safer for you to go—the top floor of a building with many windows or a basement with no windows. Explain your answer.

Critical Thinking

1. Infer how you take in water with every breath.

2. Compare and contrast four different kinds of clouds.

3. Based on what you have learned about hurricanes, **draw a conclusion** about why a hurricane will die out when it moves over land.

4. Generalize why you should not water plants with water from the ocean.

Unit C Review

Reviewing Words and Concepts

Choose at least three words from the Chapter 1 list below. Use the words to write a paragraph about how these concepts are related. Do the same for each of the other chapters.

Chapter 1
erupt
landform
lava
plain
plateau
volcano

Chapter 2
conserve
fuel
landfill
mineral
natural resource
recycle

Chapter 3
axis
phase
revolution
rotate
satellite
tide

Chapter 4
blizzard
cloud
condense
tornado
water cycle
water vapor

Reviewing Main Ideas

Each of the statements below is false. Change the underlined word or words to make each statement true.

1. A <u>plateau</u> is a mountain that forms from hardened lava.

2. A <u>magma</u> is a shape on the earth's surface.

3. A material that forms in the earth from nonliving matter is a <u>fossil</u>.

4. A <u>mineral</u> is a material that produces useful heat.

5. <u>Ore</u> is made of decayed organisms in soil.

6. The sun is a <u>planet</u>.

7. Earth <u>revolves</u> on its axis.

8. All the different shapes of the moon together are called the <u>orbits</u> of the moon.

9. Water that falls from clouds to the ground is called <u>water vapor</u>.

10. The process of using and reusing water is called the <u>rain gauge</u>.

Interpreting Data

The following models show the amount of rain collected from four rain gauges. Each line on the rain gauges is 1 centimeter. Use the models to answer the questions below.

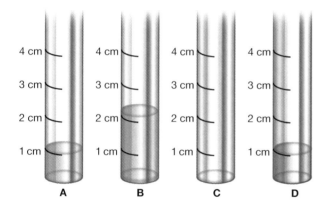

1. Which rain gauge collected no rainfall?

2. Which rain gauge collected the most rainfall? How much rain did this rain gauge collect?

3. Which rain gauges collected the same amount of rainfall?

4. What is the total amount of rainfall collected by all four rain gauges?

Communicating Science

1. Draw and label before and after pictures that show how plants, animals, and people change the surface of the earth.

2. Draw and label a diagram that shows all the things that make up soil.

3. Draw a picture that shows what causes day and night.

4. Draw and label a diagram that shows how storms form.

Applying Science

1. Write one or two paragraphs about the history of an aluminum can. You might also use pictures to tell the story.

2. You turn on the faucet to pour yourself a glass of water. Explain how the water from your faucet might have once been in the ocean.

Science Visitor Center

Using what you learned in this unit, complete one or more of the following activities to be included in a Science Visitor Center. These exhibits will help visitors learn more about the earth and the solar system. You may work by yourself or in a group.

Planets

Choose a planet in the solar system. Make a travel poster or bulletin board display that tells people why they should visit that planet.

Natural Resources

Talk to members of your family, your neighbors, and your friends. Find out how they conserve natural resources. Make a chart that shows the ways that they reuse or recycle natural resources.

Volcanoes and Earthquakes

Prepare a television news report describing either a volcano eruption or an earthquake. Tell your viewers why this event occurred. Also tell them what they can do in the future to prepare for another eruption or earthquake. You might use a video camera to tape your report.

Storms

Suppose tornadoes, hurricanes, or blizzards occur where you live. How might you keep yourself safe? How would you help others stay safe? Tell your plan to the visitors at the Science Visitor Center.

Erosion

Make a model of a beach using an aluminum foil pan, sand, and water. Show how erosion changes your beach. Repair your beach and show ways you can protect it from erosion.

Using Reference Sources

You can use many different reference sources to do research. For example, to find the meaning of a word, you might look in a glossary or a dictionary. When you want to know where a place is located, you might look at an atlas. If you want to learn more about something you studied in school, you might read a book or a magazine. You might also look in an encyclopedia or do an on-line search.

Make a List

In Chapter 2, you learned about some ways people can protect natural resources. Do research to find more information about reusing, recycling, and reducing the use of natural resources. List resources that can be protected by reducing, reusing, or recycling. Identify the reference sources you used to make your list.

Create a Poster

Use your list to think of things you can do to help protect natural resources. Share your ideas with others by making a poster. You can draw the pictures yourself, download and print images from the Internet, or cut out pictures from magazines.

Remember to:

1. **Prewrite** Organize your thoughts before you write.

2. **Draft** Make a list and create your poster.

3. **Revise** Share your work and then make changes.

4. **Edit** Proofread for mistakes and fix them.

5. **Publish** Share your poster with your class.

Unit D
Human Body

Science and Technology
In Your World!

Superman Isn't the Only One with X-Ray Vision!

Since X rays were discovered about 100 years ago, doctors have used them to look inside patients' bodies. But old-fashioned X-ray pictures have a flaw: They're flat, and most body parts aren't! Today, machines can beam X rays into the body at many different angles, making a series of flat pictures. A computer, like those used to make computer-animated movies, puts the flat pictures together to form a three-dimensional picture. As the picture slowly spins on the computer screen, doctors can study an entire body part, such as a bone. You will learn about bones and other body parts in **Chapter 1 The Body's Systems.**

Robots Test New Medicines!

Robots help scientists make and test thousands of substances in their search for new medicines. Following scientists' directions, some robots can make and test hundreds of substances in a single day. Powerful computers quickly analyze the results and help scientists decide what to try next. Rapid testing means that new medicines can reach your drugstore faster than ever before. You will learn about medicines in **Chapter 2 Staying Healthy.**

What's Inside?

X rays and other special pictures help doctors look inside you. But you don't have to be a doctor to learn how your body works. Just turn the page!

lung

lung

heart

liver

stomach

large intestine

small intestine

Chapter 1
The Body's Systems

Lesson 1
What Parts Make Up Your Body?

Lesson 2
How Do Bones and Muscles Work?

Lesson 3
What Are Some Other Body Systems?

Inquiring about **The Body's Systems**

What are the body's systems?

What parts make up the body's systems?

What are organs and tissues made of?

What is the job of the skeletal system?

How does a broken bone heal?

How do joints help you move?

What do muscles do?

How do muscles help you move?

What are the different kinds of muscles?

Why are your heart and blood vessels important?

What jobs do your brain and nerves do?

Why are your lungs and breathing important?

What do your stomach and intestines do?

Copy the chapter graphic organizer onto your own paper. This organizer shows you what the whole chapter is all about. As you read the lessons and do the activities, look for answers to the questions and write them on your organizer.

Exploring Balance

Process Skills

• observing
• inferring

Materials

• clock with a second hand

Explore

1 You will be testing your balance. Your partners will stand by to make sure you don't fall. As you do the test, **observe** how your muscles work to help you keep your balance.

2 Stand with your feet together and with your arms out. Have another student time how long you can stand with one foot off the floor. Put your foot down when needed to avoid falling. Record the time.

3 Repeat the test with your arms out and eyes closed.

Reflect

1. In which test did you keep your balance longer?

2. How did your muscles work to help you keep your balance?

3. Make an **inference.** What are some parts of your body that help you keep your balance?

? Inquire Further

Can you improve the time you can balance on one foot by practicing? Develop a plan to answer this or other questions you may have.

Using Graphic Sources

When you read a book, you might get some information from **graphic sources**, which are pictures, diagrams, or graphic organizers. These items can help you understand or organize the material. They also might show things that you would not be able to see. Most pictures in this book also have a **caption** that helps explain the picture. As you read the following chapter, *The Body's Systems,* think about the information you find in the graphic sources.

Reading Vocabulary

graphic sources (graf′ik sôrs′iz), pictures or diagrams that give information

caption (kap′shən), written material that helps explain a picture or diagram

Example

The picture is part of the photo on pages D8-D9. Part of its caption says, "Body systems work as a team to support life." The chart below lists some other graphic sources in Lesson 1, "What Parts Make Up Your Body?" Make a chart like this one on your own paper. As you read Lesson 1, look at the graphic sources. Write a sentence from each caption that connects the picture with the text.

Graphic Sources	Sentence from Caption
1. page D11: Body Systems	
2. page D12: Microscope	
3. page D13: Cells and tissues	

Talk About It!

1. How can graphic sources give information?

2. What graphic sources can you make to help you understand and organize information?

You will learn:

- about your body's systems.
- what parts make up the body's systems.
- what makes up organs and tissues.

Glossary

system (sis′təm), a group of body parts that work together to perform a job

Lesson 1

What Parts Make Up Your Body?

A quick pass brings the ball to you. You kick. You score! Of course, you didn't win this soccer game by yourself. All the players on your team helped. The parts of your body work as a team too.

The Body's Systems

A **system** is a group of body parts that work together to perform a job. Several different systems make up the human body. Like each player on a team, each system of your body has its own special task to do.

On a soccer team such as the one shown here, some players score points. Other players defend the goal. In your body, one system allows you to breathe in and out. Another system turns the food you eat into fuel that your body can use. A different system carries blood through your body. Other systems hold you upright, help you move about, and allow you to read and understand the words in this book.

To win games, soccer players must work as a team. Your body systems work together too. Every minute of every day, the different systems of your body work as a team to keep you alive and healthy.

A group of players work as a team to win games. Body systems work as a team to support life. ▼

Glossary

Glossary

organ (ôr'gən), a body part that does a special job within a body system

Parts That Make Up Body Systems

You probably know that you have a heart, a stomach, and two lungs. Each of those body parts is an **organ.** If you could see inside your body, you would have no trouble telling the organs apart. Your heart looks very different from your stomach, your lungs, and all of your other organs. Your heart also does its own special job of pumping blood. No other organ does that. Each organ in your body has its own special job.

Choose one of the body systems shown in the students' drawings. Tell everything you already know about that system and about the organs that make it up. ▶

Digestive system

Nervous system

Skeletal system

Each of your body systems is made of organs. Your heart is an organ of the circulatory system. Your stomach is an organ of the digestive system. Your lungs are organs of the respiratory system.

Other body systems include the skeletal system, muscular system, and nervous system. Organs in each system work together and depend on each other. You can see six body systems on these pages.

Circulatory system

Muscular system

Respiratory system

Glossary

tissue (tish′ü), a group of cells that look alike and work together to do a certain job

cell (sel), the basic unit of all living things, including the human body

Parts That Make Up Organs and Tissues

Each organ in your body is made of two or more kinds of tissue. A **tissue** is a group of cells that look alike and work together to do a certain job. **Cells** are the basic units, or building blocks, of the human body. In fact, all living things are made of cells.

Your body has millions and millions of cells. Cells are tiny. To see a cell, you must look through a microscope like the one on this page.

You have hundreds of different kinds of cells in your body. Muscle cells, nerve cells, and bone cells are three examples. Find the pictures of those cells on the next page. Notice that each has a different shape.

A microscope is a special tool that makes things look larger than they really are. Microscopes help scientists and students study tiny things such as cells. ▶

Muscle cells form muscle tissue. Your muscles change their shape to help you move about.

Nerve cells form nerve tissue. Your brain and nerves are made of nerve tissue. Your brain controls what you think and do. It gets and sends messages along nerves.

Bone cells form bone tissue. Your bones hold you up. Bones also work with muscles to help you move. Some bone cells also produce blood cells.

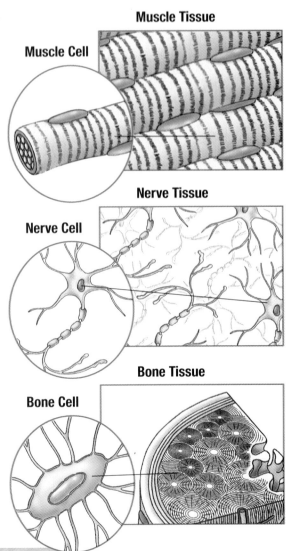

Muscle Tissue

Muscle Cell

Nerve Tissue

Nerve Cell

Bone Tissue

Bone Cell

▲ *Muscle cells and muscle tissue look different from nerve cells and nerve tissue, which look different from bone cells and bone tissue. Each kind of cell and tissue has a certain job to do.*

Lesson 1 Review

1. Tell two jobs that body systems do.

2. Name one organ in each of these body systems: circulatory system, digestive system, respiratory system.

3. What are organs made of, and what are tissues made of?

4. **Graphic Sources**
   Choose one of the groups of cells shown on this page. Make a diagram that shows how the cells are related to tissues, organs, and systems.

You will learn:
- about the job of the skeletal system.
- how a broken bone can heal.
- how your joints help you move.
- what muscles do.
- how your muscles help you move.
- about different kinds of muscles.

Lesson 2

How Do Bones and Muscles Work?

What can a hand puppet do when there is no hand inside it? Not much. It's so floppy that it collapses in a heap. Without bones and muscles, your body would be something like a floppy puppet.

The Skeletal System

About two hundred bones make up your body's skeleton, also called the skeletal system. Your skeleton does for your body what a hand does for a puppet like the one in the picture.

The bones of your skeleton help give shape to your body. Your bones also support you. Place a hand on one hip. The bone you feel is part of your pelvis. The bowl-shaped pelvis helps support your upper body when you sit and stand. Bones also work with muscles to help you move in all the ways that you do.

◀ *This puppet doesn't just flop around! Its "skeleton" is the hand inside it.*

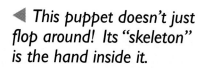

Parts of the skeleton also protect your soft organs. Touch the top of your head. You can feel your skull, which protects your brain like a built-in helmet. Touch the ribs along your sides. The ribs form a cage that protects your heart and lungs.

Bones come in many shapes and sizes, as the picture of a skeleton shows. Bones are hard but not solid. They weigh less than you might expect. The longest and thickest bone in your body is the one between your hip and knee. Its strength helps hold you upright as you run and kick.

When you were born, your skeleton was made mostly of a rubbery tissue called **cartilage**. As you grow, bone replaces most of the cartilage. However, some cartilage remains. For example, your ears and the tip of your nose are made of cartilage. That is why you can bend them.

Glossary

cartilage (kär′tl ij), a tough, rubbery tissue that makes up parts of the skeleton

Glossary

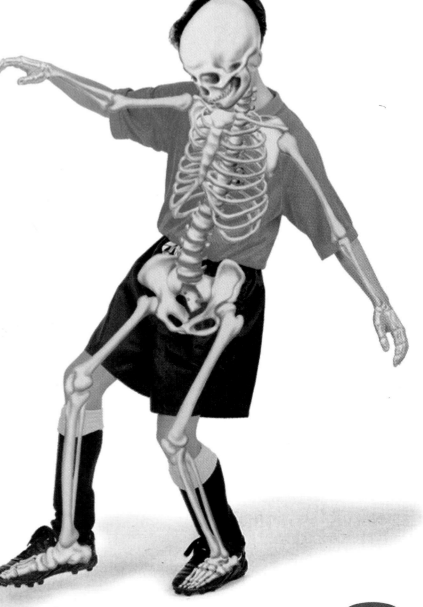

Life isn't all fun and games for your skeleton. It works hard to hold you up and help you move. ▼

How a Broken Bone Heals

Bones are strong. However, accidents can cause bones to break. Pictures called X rays help doctors see broken bones. Notice the X-ray picture of the broken bone.

Because it is living tissue, a broken bone can heal. A doctor can help the bone heal correctly by placing the injured part in a cast like the one shown. A cast holds a broken bone in the correct position while it mends.

Bone starts to mend soon after a break occurs. New bone cells begin to form. In a few days, spongy bone tissue fills the space between the broken ends of the bone. In the weeks that follow, the spongy tissue hardens until the bone is completely mended.

▲ The top X ray shows a bone broken below the knee. The bottom X ray shows the same bone after it mended.

A cast can be annoying, but it's necessary to help a broken bone heal correctly. ▶

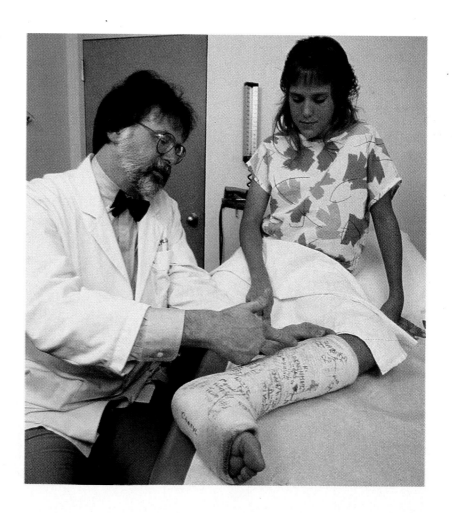

How Joints Help You Move

The place where two bones of the skeleton come together is called a **joint**. By themselves, bones are stiff. Joints allow movement. They give your skeleton its ability to bend, twist, and turn.

You have hinge joints in your elbows and knees. Find the pictures of those joints. Hinge joints work like door hinges. They let your arms and legs bend in one direction.

A layer of cartilage covers the ends of bones that meet at joints. The cartilage keeps the bones from grinding together and wearing out. Tissues called **ligaments** hold bones together at joints. You can see the ligaments of a knee joint below.

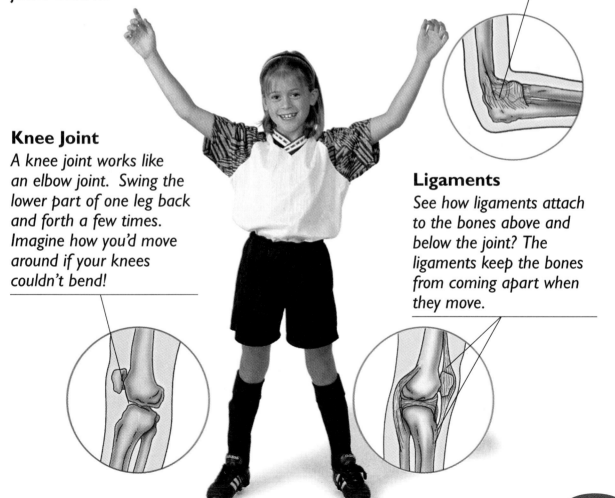

Elbow Joint

Hold one arm out. Swing the lower part away from you. Now swing it toward you. See how your elbow is like a door hinge? Open! Close! Open! Close!

Knee Joint

A knee joint works like an elbow joint. Swing the lower part of one leg back and forth a few times. Imagine how you'd move around if your knees couldn't bend!

Ligaments

See how ligaments attach to the bones above and below the joint? The ligaments keep the bones from coming apart when they move.

Glossary

muscle (mus′əl), body tissue that moves parts of the body

tendon (ten′dən), a strong cord of tissue that attaches a muscle to a bone

What Muscles Do

A **muscle** is a tissue made of muscle cells. Muscles have the job of making other body parts move. Muscles also help give your body its shape and help protect the soft organs inside you.

Your body's muscular system includes more than six hundred muscles. Most of those muscles move bones, allowing you to walk, lift, or kick a ball as the boy on the next page is doing.

Some muscles move body parts that are not bones. The muscles that move your eyebrows and lips help you smile, frown, or make a funny face. You can speak and sing because muscles move your lips, tongue, and lower jawbone.

Muscles are attached to bones by cords called **tendons.** You can feel a tendon at the back of each ankle. Find that tendon in the drawing on the next page.

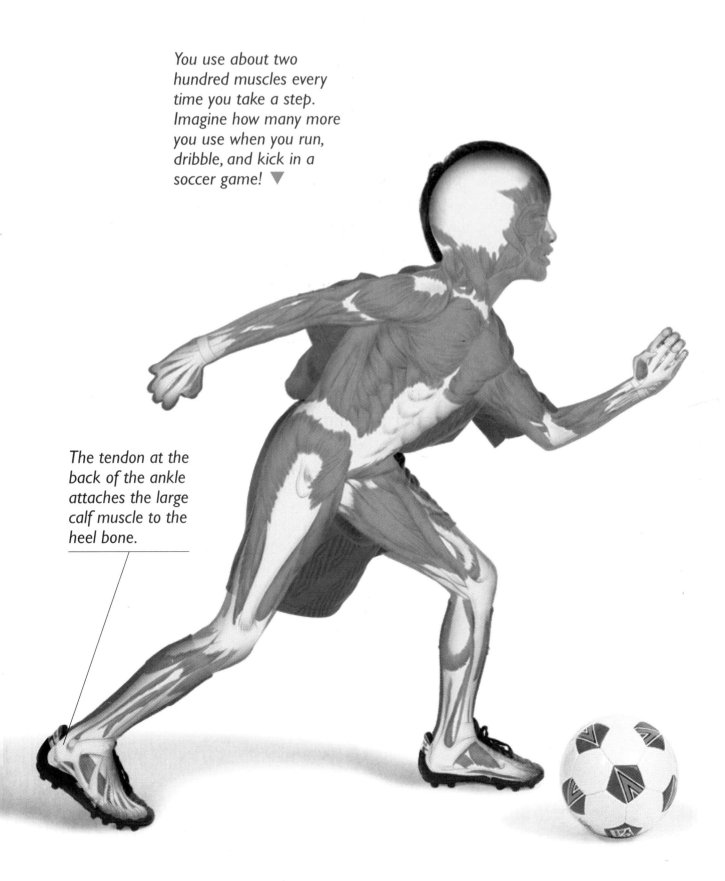

You use about two hundred muscles every time you take a step. Imagine how many more you use when you run, dribble, and kick in a soccer game! ▼

The tendon at the back of the ankle attaches the large calf muscle to the heel bone.

How Muscles Help You Move

Muscles can move bones because muscle cells can change their shape. Make a fist and lift it toward your shoulder. Notice how the muscle in your upper arm feels thick and hard. That is because the muscle cells contracted, or got shorter, to make the movement.

Muscles cannot push bones. Muscles can only pull. That is why muscles often work in pairs. One muscle pulls a bone one way. Another muscle pulls the bone the opposite way.

Look at the muscles in the picture. The top muscle in each upper arm is called the biceps. The muscle opposite the biceps is the triceps. When the biceps contracts, it pulls the boy's lower arm toward his shoulder. His arm bends. When the triceps contracts, it pulls the boy's lower arm away from his shoulder. His arm then straightens out.

Describe how the biceps and triceps of each arm look. ▼

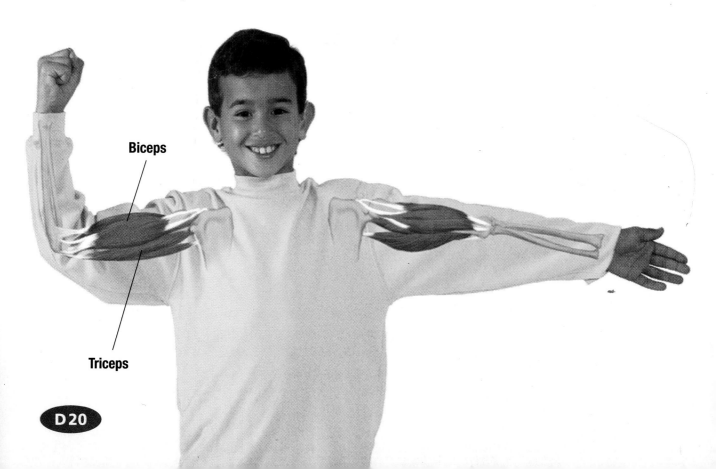

Biceps

Triceps

Different Kinds of Muscles

Arm muscles are examples of **voluntary muscles.** You can control what they do and when they do it. Certain other muscles are **involuntary muscles.** They work without your control. They work even when you sleep. For example, the muscles that move food through your digestive system are involuntary muscles.

Your heart is a special kind of involuntary muscle. It looks something like a voluntary muscle and something like an involuntary muscle. However, it works without your control.

Some people have difficulty controlling their voluntary muscles. They may need special treatment or special tools, such as a wheelchair, to help them.

Glossary

voluntary
(vol′ən ter′ē)
muscle, the kind of muscle that a person can control

involuntary
(in vol′ən ter′ē)
muscle, the kind of muscle that works without a person's control

Glossary

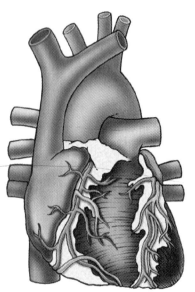

▲ *Your heart is the strongest muscle you have. It pumps blood through your body every single minute of every day.*

Lesson 2 Review

1. Tell two ways that the skeletal system helps the body.

2. Describe how a broken bone heals.

3. How do joints help you move?

4. What are two things that muscles do?

5. How do muscles help you move?

6. What are the differences between two kinds of muscles?

7. **Main Idea**
   Why is it important that your heartbeat and breathing are controlled by involuntary muscles?

Modeling How Muscles Work

Process Skills

- making and using models
- observing
- estimating and measuring

Materials

- safety goggles
- posterboard with shapes
- scissors
- hole punch
- sharpened pencil
- brass fastener
- 2 strings
- masking tape
- metric ruler

Getting Ready

Many muscles in your body work in pairs. In this activity you can demonstrate the way muscles work by making a model of muscles that move your foot.

Follow This Procedure

1 Make a chart like the one shown. Use your chart to record your observations and measurements.

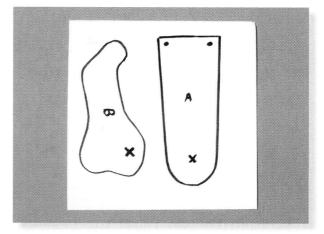

Photo A

2 Put on your safety goggles. Use scissors to cut out the shapes on the posterboard (Photo A). Use a hole punch to punch two holes at the top of shape A.

3 Use a pencil to poke a hole through the X on each shape. Place shape B on top of shape A. Attach the two pieces with a brass fastener at letter X (Photo B).

	Direction that foot moves	Length of pulled string	Length of opposite string
String near toe pulled			
String near heel pulled			

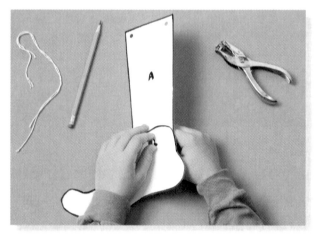

Photo B

 Safety Note *Be careful when using pointed objects.*

④ Put a piece of string through each hole. Tape down one end of each piece of string as shown (Photo C). You have made a **model** of a foot and lower leg.

⑤ Pull the string that is closer to the toe of the foot. How does the foot move? **Measure** each string from the punched hole to where the string is taped. Record your **observations** and measurements.

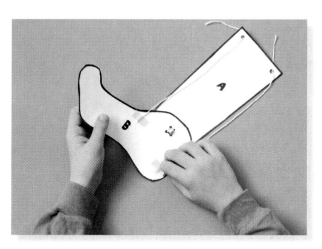

Photo C

⑥ Repeat step 5, but pull the string closest to the heel.

Interpret Your Results

1. What do the strings in the model represent?

2. What happened to the measured length of one string when you pulled on the opposite string? Describe how the strings act together to move the foot.

3. Compare and contrast your model to a real foot and leg.

 **Inquire Further**

What actually happens to your leg muscles when you stand on your toes or stand back on your heels? Develop a plan to answer this or other questions you may have.

Self-Assessment

- I followed instructions to make a **model** of a foot and leg.
- I **observed** and **measured** how the strings in the model changed as the foot moved.
- I recorded my observations and measurements.
- I described the movement of the strings and how they affected the model.
- I compared and contrasted the model foot and leg to a real foot and leg.

You will learn:

- what your heart and blood vessels do.
- about your brain and nerves.
- about your lungs and breathing.
- what your stomach and intestines do.

Lesson 3

What Are Some Other Body Systems?

Most days, you pay little attention to how your body works. Today, though, you played a hard soccer game. Your lungs took in big gulps of air. Your heart raced. What else happened inside you, and why?

Heart and Blood Vessels

Your heart is a muscular organ that pumps blood. The blood travels to all parts of your body through tubes called blood vessels. Find the heart and blood vessels in the drawing. Your heart and blood vessels make up your circulatory system.

Blood brings oxygen and nutrients to cells. When you exercise, as the boy in the picture just did, your muscle cells need extra oxygen and nutrients. That is why your heart sometimes pumps fast.

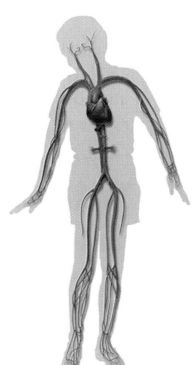

When you rest, your heart beats about ninety times a minute. It beats faster when you exercise. ▶

Brain and Nerves

You probably know that your brain controls your thoughts and feelings. Your brain also controls your actions, whether you're playing soccer or painting pictures as the boy below is doing. In fact, your brain controls heartbeat, breathing, and every other job your body does. Your brain works with your nerves. Together, the brain and nerves make up the nervous system.

You can see in the drawing that nerves thread their way throughout the body. Nerves carry messages between all parts of your body and your brain. If you need to take action, your brain sends a message along nerves to the correct muscles.

Your nervous system helps keep you safe. Suppose you touch a hot stove. Instantly, your nervous system signals your arm muscles to jerk your hand away. This fast action happens without your having to think about it.

▲ *Some nerves are thinner than a hair. Other nerves are thick. The long, very thick bundle of nerves that goes from the brain down the back is called the spinal cord.*

Your brain makes it possible for you to speak, read, write, remember, imagine, solve problems, and do hobbies such as drawing and painting. ▶

Lungs and Breathing

Your lungs and the tubes leading to them make up your respiratory system, shown below. Air enters your lungs each time you breathe in. Your body cells need oxygen from the air to stay alive. You breathe fast when you exercise because your muscle cells need extra oxygen. What other times do you breathe in an unusual way? (Hint: Look at the picture on the left.)

How does oxygen get from the air you breathe to your body cells? Oxygen in your lungs passes into your blood. Your heart pumps the blood to body cells. The blood delivers oxygen to the cells. The blood also picks up wastes that the cells have made, including a gas called carbon dioxide.

After the blood delivers its oxygen and picks up carbon dioxide, the blood is pumped back to your lungs. There, the blood gets rid of the carbon dioxide and picks up more oxygen. The carbon dioxide leaves your body when you breathe out.

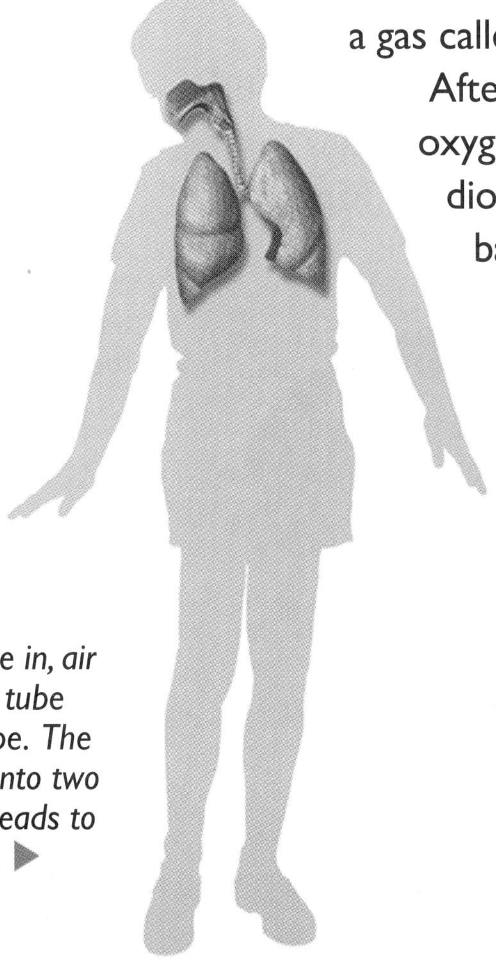

▲ Whoosh! The air you breathe out has more carbon dioxide and less oxygen than the air you breathe in.

When you breathe in, air goes down a long tube called the windpipe. The windpipe divides into two tubes. One tube leads to each spongy lung. ▶

You may think you eat because you're hungry. Actually, you eat to stay alive, active, and healthy. Your digestive system changes food so that your cells can use it. ▶

Stomach and Intestines

A tasty meal like the one shown must be changed into a form that body cells can use for fuel. Your digestive system does this job. The system includes your stomach and intestines, shown below, which help break down food.

In your stomach, food is churned and mixed until it forms a liquid. In your small intestine, nutrients from the liquid pass into your blood. Your heart pumps the nutrient-rich blood to body cells. The parts of food that your body cannot use go into your large intestine. Later, these wastes leave your body.

Lesson 3 Review

1. What two important things does blood carry to body cells?

2. How is your brain connected with the rest of your body?

3. How does oxygen that is in the air get into your blood?

4. What happens to food in your stomach?

5. **Graphic Sources**
   Look at the caption for the top picture on page D 25. What information does this caption give?

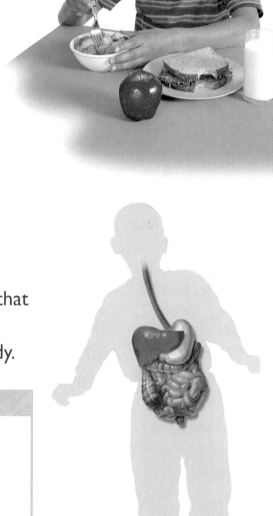

▲ *Food goes down a long tube to reach your stomach. Your intestines also are tubes, all curled up. The small intestine is longer, but narrower, than the large intestine.*

Observing Your Nervous System in Action

Process Skills

- observing
- inferring

Materials

- safety goggles
- half-meter stick

Getting Ready

In this activity you can find out more about your nervous system by testing your reaction time.

Follow This Procedure

1 Put on your safety goggles. Make a chart like the one shown. Use your chart to record your data.

	Distance half-meter stick fell
Trial 1	
Trial 2	
Trial 3	
Average distance	

2 Have a partner hold the half-meter stick vertically with the 0 end at the bottom. Hold your hands just below the stick as shown (Photo A).

3 Look closely at the half-meter stick as a partner gets ready to drop it.

Photo A

Photo B

4 When your partner drops the half-meter stick, catch it as quickly as you can. **Observe** how far the stick falls before you catch it by reading the number at the top of your hand (Photo B). Record the number in your chart.

5 Repeat steps 3 and 4 two more times.

6 Add the distances together and divide the total by 3 to get the average distance. Record your calculations.

Self-Monitoring
Did I record all my data and calculations?

Interpret Your Results

1. What was the average distance the half-meter stick fell before you caught it?

2. Look at the results for each trial. The lower the number, the faster is your reaction time. Which trial was your fastest?

3. Make an **inference.** What did your nervous system have to do from the moment your partner dropped the half-meter stick to the moment you caught it?

**Inquire Further**

Will your average reaction time become faster with practice? Develop a plan to answer this or other questions you may have.

Self-Assessment

- I followed instructions to **observe** how the nervous system works.
- I observed how far the half-meter stick fell before I caught it.
- I recorded my observations.
- I calculated the average distance the half-meter stick fell.
- I made an **inference** about how my nervous system works.

Chapter 1 Review

Chapter Main Ideas

Lesson 1
• Each body system has its own special job, but all body systems work together to support life.
• Each body system is made of organs, which work together and depend on each other.
• Organs are made of tissues, and tissues are made of cells, which are the building blocks of the body.

Lesson 2
• Bones help give shape to the body, support the body, help the body move, and protect organs.
• A broken bone heals by forming new bone cells and tissue.
• Bones come together at joints, allowing the skeleton to move.
• Muscles make body parts move, help give shape to the body, and help protect organs.
• Muscles move bones by changing shape and pulling on the bones.
• The body has both voluntary muscles and involuntary muscles.

Lesson 3
• The heart pumps blood through blood vessels to all parts of the body.
• The brain works with nerves to control thoughts and actions.

• The lungs take in air, which contains oxygen that the body needs, and also get rid of wastes.
• The stomach and small intestine change food into nutrients.

Reviewing Science Words and Concepts

Write the letter of the word or phrase that best completes each sentence.

a. cartilage
b. cell
c. involuntary muscle
d. joint
e. ligament
f. muscle
g. organ
h. system
i. tendon
j. tissue
k. voluntary muscle

1. A body part that does a special job within a body system is an ____.

2. The kind of tissue formed by muscle cells is called ____.

3. A ____ is a flexible tissue that holds bones together at a joint.

4. The ____ is the basic unit of all living things.

5. A group of cells that look alike and work together is a ____.

6. The kind of muscle that a person can control is a ___.

7. A ___ is a group of organs that work together to perform a job.

8. An ___ is a kind of muscle that works without a person's control.

9. A ___ is a strong cord that attaches a muscle to a bone.

10. The ears and tip of the nose are made of a tissue called ___.

11. The place in the elbow where bones come together is a ___.

Explaining Science

Draw and label a picture or write a paragraph to answer these questions.

1. What makes up each of the following: a tissue, an organ, and a system?

2. How do the joint in your elbow and the muscles in your upper arm help you bend your arm?

3. Choose one body system: circulatory, nervous, respiratory, or digestive. What are the main parts of the system, and what job or jobs do they do?

Using Skills

1. Use the **graphic sources** below to answer the questions. Which picture shows a nerve cell and which shows nerve tissue? How can you tell?

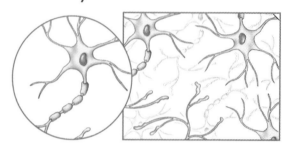

2. Sit with both knees bent and your feet flat on the floor. Straighten one leg so that it is sticking out in front of you. Then bend the leg so that it returns to where it was. What do you **infer** is happening with the muscles in your upper leg as you straighten and bend your leg?

Critical Thinking

1. Your six-year-old neighbor broke his leg a week ago. Now he's tired of wearing the cast and wants to take it off. He comes to you for help. **Apply** what you've learned about broken bones to this situation. Write what you will say to your young neighbor.

2. Make a **generalization.** When you are asleep, does your heart beat faster or more slowly than when you are awake? Explain your reasoning.

Good News!

It's exciting news when scientists invent a medicine. Here's some news that's even better: Most ways to stay healthy are a lot more fun than taking medicine!

Finish Line

Chapter 2
Staying Healthy

Inquiring about Staying Healthy

Lesson 1
What Are Some Ways to Stay Healthy?

- How do nutrients help your body?
- How can you use the Food Guide Pyramid?
- How does exercise help your body?
- How can you exercise safely?
- Why are rest and sleep important?

Lesson 2
What Are Germs?

- What harm can some germs do?
- How does the body fight disease germs?
- How can you keep germs from spreading?

Lesson 3
How Do Some Substances Affect the Body?

- How can people use medicines safely?
- What are some harmful effects of alcohol?
- What are some harmful effects of tobacco?
- What are some harmful effects of illegal drugs?

Copy the chapter graphic organizer onto your own paper. This organizer shows you what the whole chapter is all about. As you read the lessons and do the activities, look for answers to the questions and write them on your organizer.

Exploring Food Choices

Process Skills
- classifying
- predicting

Process Skills

Materials
- pencil or marker
- paper

Explore

❶ For two days, keep a "food diary" in which you list everything that you eat and drink each day. Include amounts, such as the number of bread slices you eat.

❷ **Classify** the items from your food diary. Your teacher will give you the names of six groups. Make a chart by writing those names across the top of a large piece of paper. Write each item from your diary under the correct group name as shown.

Reflect

1. In which groups did you have the most items? the fewest items?

2. In this chapter you will learn about foods needed for good health. Make a **prediction**. Do you think your food choices will be considered healthy choices? Check your prediction as you learn about healthy food choices.

? Inquire Further

Do students your age make healthier food choices than older students do? Develop a plan to answer this or other questions you may have.

Making Pictographs

This tally table shows students' votes for their favorite breakfast foods. Make a **pictograph** to make the data easier to read.

Materials
• grid paper

Math Vocabulary

pictograph
(pik′tə graf), a graph that uses pictures or symbols to show data

Remember

The key tells you what each symbol shows.

Our Favorite Foods for Breakfast		
Food	**Tally**	**Number**
Fruit	II	2
Rice	IIII	4
Bagels	HHT HHT IIII	14
Cereal	HHT HHT HHT HHT HHT HHT HHT HHT II	42
Eggs	HHT I	6
Pancakes	IIII	4

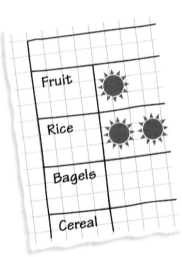

Work Together

1. Use grid paper to make a pictograph.

2. Decide what symbol to use on your pictograph. Have each symbol = 2 votes. Write a key for your pictograph.

3. Make sure your pictograph has a title.

4. How many symbols will you draw to show votes for cereal? How many symbols will you draw to show votes for bagels?

5. Complete your pictograph.

Talk About It!

How did you know how many symbols to draw for each breakfast food?

▼ *Mexican chiliquillas*

▼ *South Indian dosa*

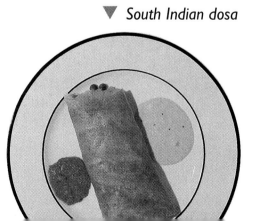

You will learn:

- how nutrients help your body.
- how to use the Food Guide Pyramid.
- how exercise helps your body.
- how you can exercise safely.
- why rest and sleep are important.

Glossary

nutrient
(nü′trē ənt), a substance in food that living things need for health and growth

Lesson 1

What Are Some Ways to Stay Healthy?

Your body does amazing things! You can breathe, talk, and move around without even thinking about it. But to work its best, your body needs to be cared for. That takes some thought!

Nutrients and the Body

The people you see on these pages have thought about how to care for their bodies. They realize that eating healthful meals is important. Food contains **nutrients** that body cells need to live and do their work. Nutrients give the body energy. Nutrients help the body grow and repair itself. Nutrients also help the body work as it should.

◀ This girl tries to eat healthful meals each day. She makes sure to drink plenty of water every day too. That's especially important in hot weather or when she is working or playing hard.

Calcium is one nutrient you may have heard of. Calcium is important for growing strong bones and teeth. Milk, cheese, and yogurt are rich sources of calcium.

Did you know that water is a nutrient? Water brings other nutrients to cells and takes wastes away from cells. Water also helps keep your body temperature steady.

Some foods, such as milk, provide many different nutrients. However, no single food can give you all the nutrients your body needs. To stay healthy, you need to eat different kinds of food each day.

Like members of a team, the nutrients in different foods work together to help the body stay healthy. Do your part by eating a variety of foods. ▼

The Food Guide Pyramid

Look around any large grocery store, and you'll see hundreds of different foods. You may wonder how you can possibly choose the foods that will give you all the nutrients you need. An easy way to plan healthful meals is to use the Food Guide Pyramid shown on these two pages. It tells you how much of each kind of food to eat each day.

All of the parts of the Food Guide Pyramid—except fats, oils, and sweets—identify different food groups. Each group contains foods that have similar nutrients.

Look at the bottom of the pyramid. Notice the suggested servings for bread, cereal, rice, and pasta. What happens to the suggested serving size as you go up the pyramid?

Milk, Yogurt, and Cheese Group
Eat 2–3 servings each day. The foods in this group have many different kinds of nutrients. Your body uses these nutrients for growth and repair, energy, and working well.

Vegetable Group
Eat 3–5 servings each day. As with the Fruit Group, most of the nutrients in vegetables help your body work as it should.

Fats, Oils, and Sweets

Eat very little of these foods. These foods are not a food group. They can make a person gain too much body fat. In addition, sugary foods can cause tooth decay. Also, fatty foods are not good for the heart and blood vessels.

Meat, Poultry, Fish, Dry Beans, Eggs, and Nuts Group

Eat 2–3 servings each day. Most of the nutrients in these foods help your body grow and repair itself when needed.

Bread, Cereal, Rice, and Pasta Group

Eat 6–11 servings each day. Most of the nutrients in the foods in this group give your body energy.

Fruit Group

Eat 2–4 servings each day. As with the Vegetable Group, most of the nutrients in fruits help your body work as it should.

POPCORN

RICE

PASTA

Exercise and the Body

Making healthful food choices is one way to help your body. The children on these two pages know another way: exercise!

Exercise helps your muscles. When you exercise regularly, the cells in your voluntary muscles get bigger. As muscle cells get bigger, the whole muscle gets bigger and stronger.

Try to exercise every day. Activities such as in-line skating and tennis are good exercise, but you also can exercise by doing chores. Exercising with a friend or family member is fun, but you don't need a partner to be active. You don't need special equipment. You don't even need to leave your house!

Your heart is a special kind of muscle. It gets stronger when you exercise regularly. The muscles that help you breathe in and out get stronger too. As a result, your lungs work better.

Taking walks and playing active games are just two ways to get the exercise you need for good health. Regular exercise helps you feel better and look better. It helps you work and play without getting too tired. It even helps you sleep better at night! What kinds of exercise do you enjoy now? What new kind of exercise would you like to try?

How to Exercise Safely

Exercise is fun, but not if you hurt a muscle or have an accident. One way to prevent muscle injuries is to warm up before exercising. Do slow stretches to get your muscles ready for exercise. After you exercise, do the same kinds of stretches to help your muscles cool down. The chart has more ideas for keeping safe when you exercise.

Some Rules for Safe Exercise

- Choose a safe place to exercise. Be sure you have enough room to move around.

- Ask a responsible adult to show you the right way to do an exercise. Some exercises can be harmful if not done correctly.

- Wear comfortable clothes that let you move freely.

- Wear shoes that support your feet and fit correctly. Rubber-soled shoes are a good choice for most sports and games.

- Keep your shoes tied. You could trip over an untied shoelace.

- Wear proper safety equipment—such as a helmet, wrist protectors, kneepads, and elbow pads—for the activity you are doing. In-line skating, skateboarding, bicycling, and softball are some activities that require safety equipment.

- Drink water before and after exercise. If you exercise for a long time, take water breaks during your exercise. This is especially important in hot weather.

Rest and Sleep

No one can be active all the time. After you work or play hard, your body needs to recover. Reading a book, as the girl in the picture likes to do, is one way to rest your body.

Sleep is a special kind of rest. Like the boy below, you need to get plenty of sleep each night. Sleep helps you grow, because your body makes new cells more quickly when you are asleep. Other body activities slow down during sleep. Because your body uses less energy when you sleep, you'll have lots of energy for the next day.

Reading, playing a quiet game, and listening to music are some ways to give your body a rest. What do you do to rest? ▼

▲ *Going to bed on time is important. You can work, play, and learn better when you get enough sleep.*

Lesson 1 Review

1. Why does your body need the nutrients in food?

2. What food group should you eat the most servings from each day?

3. What does regular exercise do for your heart?

4. Why should you warm up before you exercise?

5. Why does sleep help you grow?

6. **Make Pictographs**
   Make a pictograph showing the foods you ate today from each group in the Food Guide Pyramid.

Testing Foods for Fat

Process Skills

- observing
- inferring

Materials

- piece of paper bag
- marker
- pat of butter
- paper towel
- carrot stick
- piece of raw potato
- potato chip
- other assorted foods

Getting Ready

In this activity you can test foods to see if they contain fat.

Fat is a nutrient in many foods. People need some fat in their diets, but eating too much fat can be harmful.

Follow This Procedure

1 Make a chart like the one shown. Use your chart to record your observations.

Food	Observation of spot	Presence of fat
Butter	Light shines through	Yes
Carrot	No light shines through	No
Raw potato		
Potato chip		

2 Use the marker to draw a circle on the paper bag for each food you will test. Each circle should be the size of a large coin.

⚠️ **Safety Note** *Do not taste any of the foods, either before or after testing them.*

3 Label one circle *Butter*. Rub the butter over the circle. Then wipe your hands with a paper towel. Hold the paper up to the light (Photo A). You will see light shining through the spot. This is caused by the fat in the butter.

Photo A

Photo B

④ Label another circle *Carrot.* Rub the carrot stick over the circle (Photo B). Wipe your hands with the paper towel. If the carrot makes a wet spot, let it dry. Hold the paper up to the light. There should be no light shining through the spot because carrots do not contain fat.

⑤ Repeat step 4 for each food. Wipe your hands on the paper towel after testing each food. **Observe** any spot made by the food. Record your observations.

⑥ Record whether or not fat was present in each of the foods. Wash your hands after this activity.

Interpret Your Results

1. Which of the foods were low in fat? Which were high in fat?

2. Potato chips are made by frying potatoes in oil. Make an **inference.** How does frying foods in oil change the amount of fat in the food?

**Inquire Further**

What meals could make up a low-fat menu for a week? Develop a plan to answer this or other questions you may have.

> **Self-Assessment**
>
> • I followed instructions to test foods for fat.
> • I **observed** each food item's circle for its fat amount.
> • I recorded my observations.
> • I identified foods that are low in fat and high in fat.
> • I made an **inference** about how frying foods in oil changes the amount of fat in the food.

Glossary

Glossary

germ (jėrm), a thing too tiny to be seen without a microscope; some may cause disease

disease (də zēz′), an illness

Here you see the germs that cause colds (left) and strep throat (right). The picture of the strep-throat germs has been enlarged 8,000 times. The picture of the cold germs has been enlarged 90,000 times. Influenza (flu), pneumonia, chicken pox, measles, and mumps are some other diseases caused by germs. ▶

Lesson 2

What Are Germs?

You can't see them, but they're everywhere! They're in air, water, soil, and food. They're on everything you touch. They're even on your skin and inside your body! What are they?

Germs and Disease

If you haven't guessed, they're germs. **Germs** are tiny things—so tiny that they must be viewed through a microscope. Many germs cause no harm. However, some germs cause disease. A **disease** is an illness. Many diseases caused by germs can spread from person to person. The pictures show two kinds of disease germs.

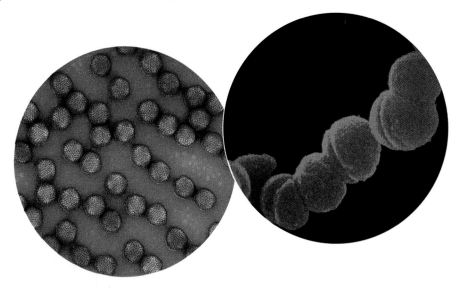

Fighting Disease Germs

To cause disease, germs must get in the body and make more germs. Your body has ways of keeping germs out. Your skin stops many germs from getting in. If you breathe germs in, hairs in your nose trap many of them. A sticky substance in your nose and throat also traps germs. You get rid of the germs when you sneeze, cough, or blow your nose.

Sometimes germs do get in. For example, a cut in the skin can let germs in. Then special blood cells go to work. Some cells surround and destroy germs. Other cells make substances that help destroy germs. Still other cells "remember" to make those substances if germs of the same kind get in the body again.

Medicines called **vaccines** can prevent certain diseases caused by germs. Some vaccines are given as shots. Other vaccines are swallowed. The child in the picture is getting a vaccine.

Glossary

vaccine (vak sēn′), a medicine that can prevent the disease caused by one kind of germ

Glossary

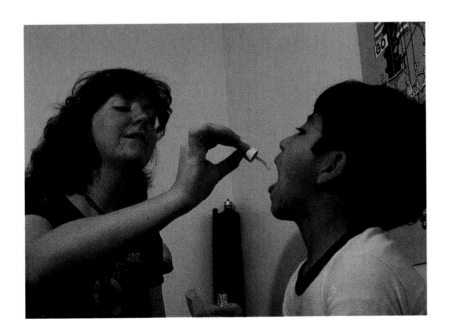

◀ Vaccines help the body fight disease germs. Each vaccine causes blood cells to "remember" to make the substances that attack one kind of germ. If that germ enters your body, the substances will attack it right away. You won't get sick. Your doctor knows what vaccines you should get.

How to Keep Germs from Spreading

If you have disease germs in or on your body, you can spread them to other people. Germs may go into the air when you cough or sneeze. Germs may get onto things such as food, dishes, and pencils. The germs may enter the bodies of other people when they breathe or put things in their mouths. To help keep germs from spreading, do what the children in the pictures do.

Wash Those Hands!

Washing gets rid of germs. When your hands are clean, you're less likely to get germs on things that others might touch or put in their mouths. You're also less likely to get other people's germs into your own body. ▼

Use Lots of Soap!

Don't just run your hands under the water for a few seconds. Wash your hands thoroughly after doing any activity that gets them dirty. Also, wash after you touch an animal, after you go to the bathroom, and before you handle or eat food. ▶

◀ **Use a Tissue!**
Cover your mouth and nose with a clean tissue whenever you cough or sneeze. Then throw the tissue away and wash your hands.

▲ **Don't Share Germy Stuff!**
Use your own drinking glass, toothbrush, towel, and washcloth. Never use someone else's fork, spoon, or dish without washing it first.

Lesson 2 Review

1. What are two diseases caused by germs?

2. Describe one way that the body fights germs.

3. Why is it important to wash your hands often?

4. **Identify the Main Idea**
Read pages D48 and D49 again. Write a sentence that tells the main idea of these two pages.

You will learn:

- how people can use medicines safely.
- some harmful effects of alcohol.
- some harmful effects of tobacco.
- some harmful effects of illegal drugs.

Lesson 3

How Do Some Substances Affect the Body?

Even if you've heard a lot about drugs, you may have questions like these: Is a medicine a drug? What's the harm in drinking or smoking? Why are some drugs against the law? You'll find the answers in this lesson.

Using Medicines Safely

The answer to the first question is yes, a medicine is a drug. A drug is a substance that causes changes in the body. Some medicines prevent disease. Other medicines help sick people, like the child in the picture, get well or feel better. However, any medicine can be harmful if used the wrong way. All medicines must be used with care.

◀ Only a responsible adult should give medicine to a child.

A **prescription medicine** is one that people need a doctor's order to buy. A pharmacist fills the order and puts a label on the medicine, like the one in the picture. Only the person whose name is on the label should take the medicine. The adult who takes or gives the medicine should follow the directions on the label exactly.

An **over-the-counter medicine** is one that people can buy without a doctor's order. It is just as important to follow the label directions on an over-the-counter medicine as on a prescription medicine. The chart has some more ideas for using medicines safely.

Glossary

prescription medicine (pri skrip′shən med′ə sən), a medicine that can be bought only with a doctor's order

over-the-counter medicine, a medicine that can be bought without a doctor's order

---

Some Rules for Safe Medicine Use

- Do not take any medicine by yourself. Take medicine only from a doctor, a nurse, or an adult responsible for you.

- Leave the labels on all medicines. That way, people know what the medicine is and what directions to follow.

- Never share prescription medicines. One person's medicine might make another person sick.

- Tell an adult responsible for you if you feel any unwanted effects from a medicine, such as an upset stomach or a headache.

- Keep medicines away from small children, and keep medicine containers closed.

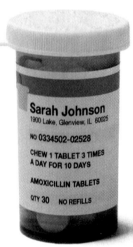

▲ Directions on a prescription medicine label may include how much medicine to take, when to take it, and how long to take it.

Glossary

alcohol (al′kə hȯl), a drug found in beer, wine, and liquor that can be harmful

Alcohol upsets the way the brain and body work. This baseball player could not perform well or be safe if she had alcohol in her body. ▼

Harmful Effects of Alcohol

Beer, wine, and liquor are drinks that contain alcohol. **Alcohol** is a drug. It can be harmful to a person's health and safety.

Alcohol goes to the brain very quickly. Alcohol changes the way a person's brain works. A person who drinks alcohol may have trouble thinking and talking clearly. The person also may walk unsteadily, feel dizzy or sleepy, and have trouble seeing clearly. A person who often drinks large amounts of alcohol can damage his or her body organs. However, the person may find it hard to stop drinking.

Alcohol makes it impossible to do many activities safely. Playing baseball is just one example of such an activity. Riding a bicycle or driving a car after drinking alcohol can lead to a serious accident.

Harmful Effects of Tobacco

Cigarettes, cigars, chewing tobacco, and pipe tobacco are made from the tobacco plant. Tobacco contains a drug called **nicotine.** Nicotine can harm the heart. Tobacco users are more likely to get heart disease than other people. Nicotine also makes it hard for a tobacco user to quit.

Tobacco smoke contains dark, sticky tar and other substances that can harm the lungs. Smokers are more likely to get lung cancer and other lung diseases than other people. Tobacco smoke also may harm the health of people who are around smokers and breathe in the smoke.

Glossary

nicotine (nik′ə tēn′), a drug in tobacco that can harm the body

Glossary

◄ *A person needs a strong heart and lungs to keep going for a long time. Smoking can harm the heart and lungs. Smokers may find it hard to do activities such as running or swimming. After a short time, they may be out of breath and too tired to continue.*

Glossary

illegal (i lē′ gəl)
drug, a drug that is
against the law to buy,
sell, or use

Harmful Effects of Illegal Drugs

Drugs that are against the law are called **illegal drugs.** These dangerous drugs include marijuana, heroin, and cocaine. Such drugs change the way the brain works. They can make a person feel very sick. They also can seriously damage body organs. Despite such effects, people who use illegal drugs usually find it hard to stop. Until a person reaches a certain age, alcohol and tobacco are also considered illegal drugs.

The students who attend the school in the picture want to stay healthy and safe. That is why they say "no" to all illegal drugs.

◀ *What does this sign tell you about this school?*

Lesson 3 Review

1. List three ways to use medicines safely.

2. What are two things that an alcohol drinker may have trouble doing?

3. What effects does nicotine have?

4. Why should people say "no" to drugs such as marijuana, heroin, and cocaine?

5. **Compare and Contrast** Compare and contrast prescription medicine and over-the-counter medicine.

Conducting a Sleep Survey

Materials

- four small pieces of paper
- numbered paper bag
- writing paper
- grid paper

Process Skills

- formulating questions and hypotheses
- identifying and controlling variables
- experimenting (survey)
- collecting and interpreting data
- communicating

State the Problem

How many hours of sleep do most third graders get?

Formulate Your Hypothesis

If you survey third graders about how much sleep they get each night, what answer will most of them give? Write your **hypothesis.**

Identify and Control the Variables

Control **variables** to conduct a fair survey. Each student must be asked the same question. Keep your answers private to avoid influencing other students' answers. Include all classmates in the survey.

Test Your Hypothesis

Follow these steps to conduct a **survey.**

1 Make a survey form like the one on the next page. Use the survey form to record your data.

2 Each student in your group should think about the following question: "How many hours of sleep do you get on a typical school night?"

3 Have each student in your group write an answer on a small piece of paper, fold the paper, and put it into the bag.

Continued ➜

④ Pass the bag to another group of students. Remove the slips of paper from the bag passed to your group. **Collect** and **record** your **data.** On your chart, make a tally mark by the number of hours of sleep written on each piece of paper. Put the slips of paper back in the bag.

⑤ Repeat step 4 until you have recorded data from all groups, including your own.

Collect Your Data

Number of hours of nightly sleep	Number of students
Less than 8	
8	
9	
10	
11	
12	
More than 12	

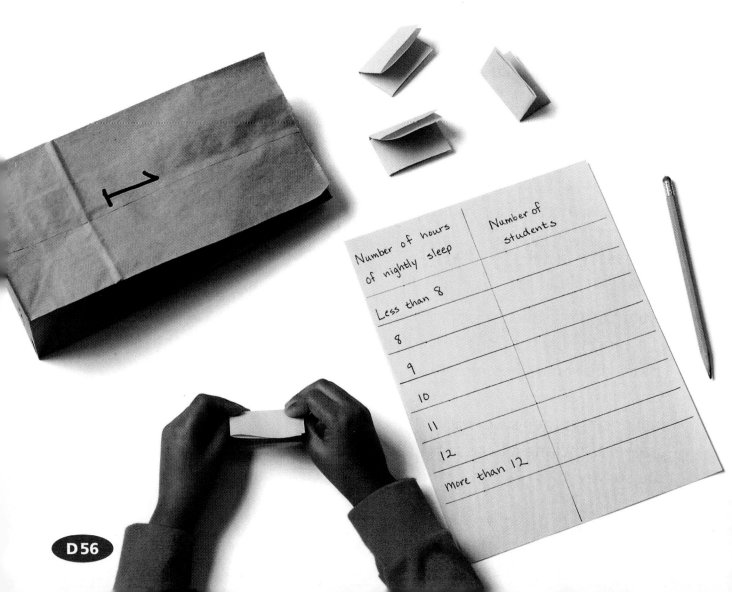

Interpret Your Data

1. Label a piece of grid paper as shown. Use the data from your chart to make a bar graph on your grid paper.

2. Study your graph. Tell how many hours of sleep the largest number of students reported.

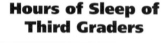

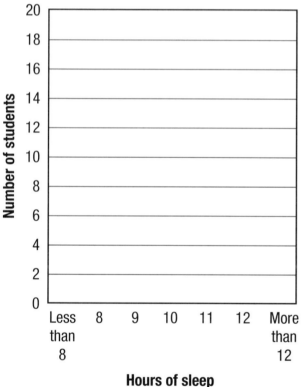

Hours of Sleep of Third Graders

Number of students: 0, 2, 4, 6, 8, 10, 12, 14, 16, 18, 20

Hours of sleep: Less than 8, 8, 9, 10, 11, 12, More than 12

State Your Conclusion

How do your results compare with your hypothesis? **Communicate** your conclusion about how much sleep most third graders get. Discuss the results with your class.

 Inquire Further

How much sleep do younger students get? Make a plan to answer this or other questions you may have.

Self-Assessment

- I made a **hypothesis** about how much sleep most third graders get.
- I **identified** and **controlled variables.**
- I conducted an **experiment** (class survey) to test my hypothesis.
- I **collected data** in a chart and **interpreted** my **data** by making and studying a graph.
- I **communicated** by stating my conclusion.

Chapter 2 Review

Chapter Main Ideas

Lesson 1

• Nutrients in food give your body energy, help your body grow and repair itself, and help your body work as it should.

• The Food Guide Pyramid tells how much of each kind of food to eat each day to get all the nutrients needed.

• Exercise strengthens the voluntary muscles, the heart muscle, and the muscles that help with breathing.

• Ways to exercise safely include warming up before exercising and cooling down after exercising.

• Sleep is a special kind of rest that helps the body grow and saves energy for the next day's activities.

Lesson 2

• Some germs cause diseases, such as colds, strep throat, flu, pneumonia, chicken pox, measles, and mumps.

• Your body has several ways to keep germs out and to destroy germs that do get inside.

• Ways to keep germs from spreading include washing hands with soap and water, covering sneezes and coughs with tissues, and not sharing such items as drinking glasses.

Lesson 3

• Ways to use medicines safely include not sharing prescription medicines and following the directions on the label of any medicine.

• Alcohol changes the way the brain works, can lead to accidents, and can damage body organs over time.

• Tobacco can harm a smoker's heart and lungs and may harm the health of other people who breathe in the smoke.

• Illegal drugs such as marijuana, heroin, and cocaine change the way the brain works, can make people very sick, and can seriously damage body organs.

Reviewing Science Words and Concepts

Write the letter of the word or phrase that best completes each sentence.

a. alcohol

b. disease

c. germ

d. illegal drug

e. nicotine

f. nutrient

g. over-the-counter medicine

h. prescription medicine

i. vaccine

1. Any medicine that requires a doctor's order is a ____.

2. A ____ is a substance in food that is needed for health and growth.

3. A medicine that can prevent the disease caused by one kind of germ is a ____.

4. Tobacco contains ____.

5. A ____ is a tiny thing that may cause disease.

6. Medicine bought without a doctor's order is ____.

7. Beer, wine, and liquor contain a drug called ____.

8. Another word for illness is ____.

9. An ____ is any drug that is against the law.

Explaining Science

Make a chart or write a paragraph to answer these questions.

1. What can you do today to stay healthy? Include food, exercise, and sleep in your answer.

2. What will your body do to fight germs today?

3. What are some reasons to stay away from alcohol, tobacco, and illegal drugs? Give at least one reason for each.

Using Skills

1. Make a **pictograph** using the data below. Decide on a symbol to use, and have each symbol equal 3 votes. Be sure to include a title and a key.

Our Favorite Kinds of Exercise		
Exercise	Tally	Number
In-line skating	✝✝✝ /	6
Bicycling	✝✝✝ ✝✝✝ ✝✝✝	15
Basketball	✝✝✝ ////	9
Soccer	✝✝✝ ✝✝✝ ✝✝✝ ✝✝✝ /	21
Jumping rope	///	3
Dancing	✝✝✝ /	6

2. Even if your favorite food is a healthful one, such as milk or apples, would it be a good idea to eat only that food? Why or why not? **Communicate** your thoughts by writing a paragraph.

Critical Thinking

1. Gayle wants Robin to try a cigarette. Robin is not sure how to refuse while still keeping Gayle as a friend. **Solve the problem.** Write a note to Robin, suggesting what she might say to Gayle.

2. The directions on an over-the-counter medicine can no longer be read. What do you **conclude** is the best thing to do with the medicine? Explain your reasoning.

Unit D Review

Reviewing Words and Concepts

Choose at least three words from the Chapter 1 list below. Use the words to write a paragraph that shows how the words are related. Do the same for Chapter 2.

Chapter 1
cartilage
joint
ligament
muscle
tendon
voluntary muscle

Chapter 2
disease
germ
illegal drug
over-the-counter
  medicine
prescription
  medicine
  vaccine

Reviewing Main Ideas

Each of the statements below is false. Change the underlined word to make each statement true.

1. A tissue is a group of <u>organs</u> that look alike and work together to do a certain job.

2. About two hundred <u>systems</u> make up the body's skeleton.

3. Muscles move bones by <u>pushing</u> them.

4. Your <u>stomach</u> works with your nerves to control your thoughts and actions.

5. The lungs and the tubes leading to them make up the <u>circulatory</u> system.

6. The body uses substances called <u>wastes</u> for energy, for growth and repair, and for working well.

7. Sleep is a special kind of <u>exercise</u>.

8. If you breathe germs in, hairs in your <u>mouth</u> trap many of them.

9. A <u>food</u> is any substance that causes changes in the body.

10. Cigarettes contain <u>alcohol</u>, which can harm the heart and make it hard to quit smoking.

Interpreting Data

Use the medicine label to answer the questions below.

1. Is this a prescription medicine or an over-the-counter medicine? How can you tell?

2. How often should this medicine be taken?

3. You are at a friend's house. Your ear starts to hurt. Your friend brings you this medicine and suggests that you take some. Why should you say no?

Communicating Science

1. Explain how body cells depend on the respiratory, digestive, circulatory, and nervous systems to stay alive. Write a paragraph.

2. Make a list of the ways that your bones help you.

3. How can a person get the nutrients that he or she needs each day? Use what you know about the Food Guide Pyramid to write an answer to that question.

4. Draw and label a picture that shows how the body keeps many germs out and fights germs that do get in.

Applying Science

1. What would your life be like if you had to think about making your involuntary muscles work? Write a diary entry for one day.

2. It's almost time for supper. Your hands look perfectly clean. Write a short paragraph that explains why you should wash your clean-looking hands before eating supper.

3. A child breaks her lower leg. When the cast finally comes off, the muscles of the lower leg are small and weak. Write a few sentences that explain why that happened and what the child can do about it.

4. Make a poster that a community group could use to discourage people from drinking alcohol.

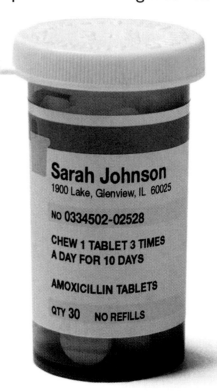

Sarah Johnson
1900 Lake, Glenview, IL 60025

NO 0334502-02528

CHEW 1 TABLET 3 TIMES
A DAY FOR 10 DAYS

AMOXICILLIN TABLETS

QTY 30   NO REFILLS

Unit D
Performance Review

Museum of the Body

Using what you learned in this unit, complete one or more of the following activities to be included in a Museum of the Body. The exhibits will help visitors learn more about body systems and staying healthy. You may work by yourself or in a group.

Body Art

Make an outline of the human body on a large sheet of paper. You might trace around a partner standing in front of paper fixed to a wall. Inside the outline, draw the circulatory system. Label the parts. If you wish, repeat for other body systems. Display your artwork in the museum.

Joint Demonstration

You've learned about hinge joints— now use your library to find out about the other joints your body has. Plan a demonstration about joints for the museum. Prepare a chart about joints to display. As you talk about each kind of joint, show the joint's movement with your own body.

A Sweet Poem

Think about your favorite kinds of fruit. Why do you like to eat them? How do they help your body? Turn your ideas into a poem titled "Fabulous Fruit." Plan to recite your poem for museum visitors. Put lots of feeling into it!

Health Pictographs

Make one or more pictographs for the museum's math room. The pictographs should show data that you have collected about your classmates' health practices. For example, you may have pictographs about how much sleep your classmates get and what kinds of exercise they do to stay healthy.

Puppet Play

The museum will have an exhibit just for children five years old and younger. Prepare a puppet play about stopping the spread of germs for this exhibit. Your puppets can be human or animal characters. You may also want to have a character named Mr. or Ms. Germ!

Using Graphic Organizers

A graphic organizer is a visual device that shows how ideas and concepts are related. Word webs, flowcharts, and tables are different kinds of graphic organizers. The graphic organizer below is an example of a flowchart. It shows how the parts of the human body you studied in Chapter 1 are related.

Make a Graphic Organizer

In Chapter 2, you learned about ways to keep the human body healthy. You also learned about substances that can harm the body. Use information in Chapter 2 to make a graphic organizer that shows what substances are harmful to the human body. Include information about how each substance is harmful.

Write a Persuasive Letter

Develop materials to convince younger children to stay away from substances that can be harmful to their bodies. Use information in your graphic organizer to write a persuasive letter or brochure for use with the children.

Remember to:

1. **Prewrite** Organize your thoughts before you write.

2. **Draft** Write your persuasive letter or brochure.

3. **Revise** Share your work and then make changes.

4. **Edit** Proofread for mistakes and fix them.

5. **Publish** Share your letter or brochure with your class.

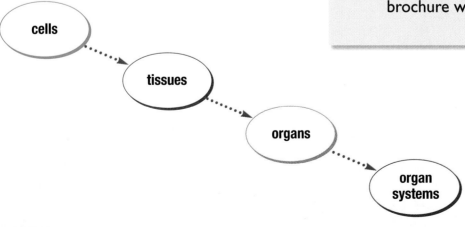

cells → tissues → organs → organ systems

Your Science Handbook

⚠️ Safety in Science

Scientists know they must work safely when doing experiments. You need to be careful when doing experiments too. The next page shows some safety tips to remember.

Safety Tips

- Read each experiment carefully.

- Wear safety goggles when needed.

- Clean up spills right away.

- Never taste or smell substances unless directed to do so by your teacher.

- Handle sharp items carefully.

- Tape sharp edges of materials.

- Handle thermometers carefully.

- Use chemicals carefully.

- Dispose of chemicals properly.

- Put materials away when you finish an experiment.

- Wash your hands after each experiment.

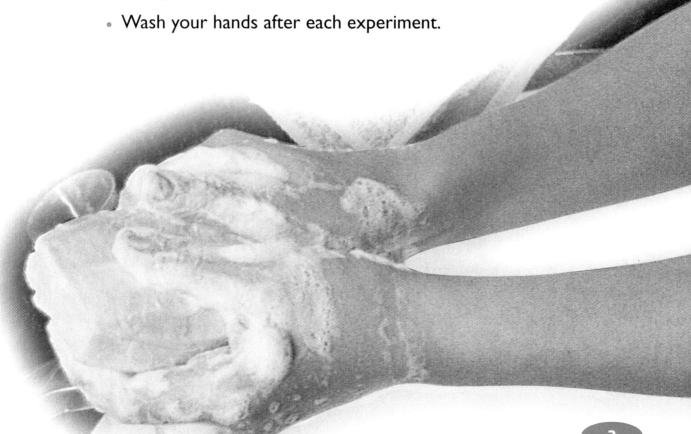

Using the Metric System

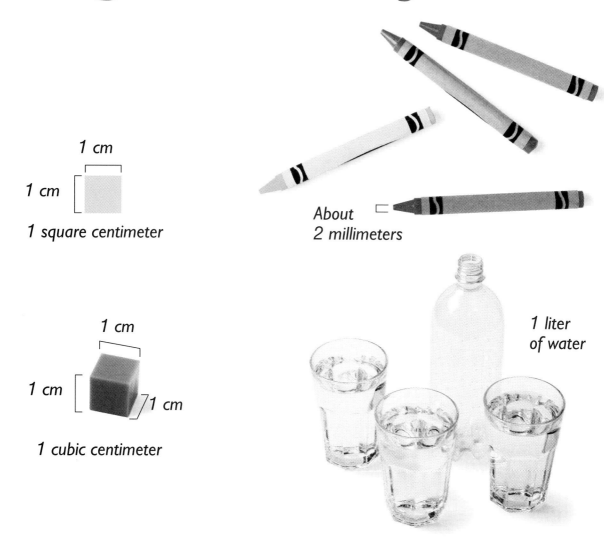

1 cm

1 cm

1 square centimeter

About
2 millimeters

1 cm

1 cm

1 cm

1 cubic centimeter

1 liter
of water

11 football fields end to end
is about 1 kilometer

About 1 centimeter

About 1 kilogram

About 1 meter

Water boils
(100° C)

Normal body
temperature
(37° C)

Water freezes
(0° C)

Observing

How can you improve your observations?

There are different ways to observe things. The method you choose can affect how fully and accurately you notice and record things. Observation is a very important part of scientific investigation.

Observation involves the use of all five senses. Effective observing requires you to use your senses of sight, hearing, smell, and touch to gather information about objects and events. Taste is very important too, but you should never taste an unknown object—it could be dangerous!

When you are first observing something, you should be quiet and still. Then you can touch things, smell them, move them, shake them gently, and compare them with others.

Organizing your senses intelligently can help you make better observations.

Practice Observing

Materials

- small hand bell

Follow This Procedure

1. Sit quietly while you observe the bell with your eyes, ears, and nose. Do not taste or touch the bell.

2. List each sense and describe what you observed with it.

3. Touch the bell with your fingers. Move it on its side to observe all of its parts.

4. Pick the bell up and shake it.

5. Compare what you observed about the bell before you shook it with what you observed afterwards. What changed? What sense did you use to observe it?

6. Compare your bell with the bell of your partner. Do you observe any difference with your eyes, ears, or nose? Can you tell the two bells apart by observing them just with your sense of hearing?

7. Compare the size of your bell with your partner's bell. Observe the length, width, and height of the bells.

Thinking About Your Thinking

List the steps you used to learn about the shape, feel, and look of the bell.

List all of the steps you used to observe what happens when the bell makes a sound. How many senses did you use?

List the things you did to compare the size of your bell and your partner's bell.

Communicating

How do you communicate clearly?

You communicate when you use words, pictures, and body language to share what you observe.

To communicate the difference between two very similar things, you need to use clear, precise communication. Some forms of communication work better than others in different situations.

Putting accurate information into a table can help you to communicate more clearly.

Practice Communicating

Materials

- two transparent plastic containers
- distilled water
- olive oil
- magnifying glass
- paper towels

Follow This Procedure

1. Work in pairs.

2. First, on your own, observe the properties of the water and describe them in your journal. Describe its color or lack of color. Is it transparent? (Can you see through it?) How well? Describe its smell. Is it wet? Is it slippery? Is it slimy? Are things floating in it? Draw a picture of the water.

3. Make a table with a column to communicate your observations about water.

4. Repeat step 2 with olive oil. Use the paper towels to wipe up spills immediately. Wash your hands after touching the oil.

5. In your table, next to your water observations, add another column to communicate your observations about the olive oil.

6. Pick one liquid in your table. Read its properties to your partner until he or she can tell which liquid you are describing.

7. Reverse roles. Now your partner can pick one liquid in his or her table. Try to determine if your partner is describing water or oil.

8. Show each other your tables and compare them.

Thinking About Your Thinking

Why were drawings not very useful for telling water and oil apart?

Did creating a table make communication easier?

Think about the importance of clear communication. Imagine if you asked someone for a glass of chocolate and you got a glass of chocolate oil instead of chocolate milk! Sometimes communication must be exact!

Classifying

How can you recognize common properties?

Classifying means arranging or grouping objects according to their common properties.

To be a good classifier, you need to learn about, understand, and recognize the properties of objects. You also need to be able to create new ways to group objects.

Objects that you classify may have some properties in common but not others. As a result, they may be classified together in some groupings and not in others.

Developing an organized way to classify objects can be very helpful. With organization and practice, your classifying skills will be right on the money.

▲ *50 pennies*

▲ *25 shiny pennies*

Practice Classifying

Materials

- 5 pennies
- 2 nickels
- 2 foreign coins
- 10 dimes
- 4 quarters

Follow This Procedure

1 Work in groups of four.

2 Name one group, or classification, to which all of the objects belong. When you begin to classify a group of objects, you might look at the big groups that they belong to. The objects in front of you could be grouped as money, coins, or metals.

3 Which coins aren't used in the United States? Classify your coins in two groups: (1) U.S. coins and (2) foreign coins.

4 What is the name of the coin that is worth one cent? What coin is worth five cents? Ten cents? Twenty-five cents? Group the coins by name and record how many you have in each classification.

5 Create a classification called "worth a dollar." How many quarters would you need to put together to be in the "worth a dollar" group? How many dimes would you need?

Thinking About Your Thinking

What classification groupings did you use to separate the coins? Which coins do you use every day?

Coin collectors, or numismatists, classify coins in many other ways. A rare penny that is in excellent, or mint, condition, is worth much more than one cent to a collector. Some are worth thousands of dollars! In what other ways might you classify coins?

Estimating and Measuring

How can you estimate and measure correctly?

An estimate is your best guess about how heavy, how long, how tall, or how hot an object is.

Once you've estimated the object's properties, you can then measure and describe them in either metric or customary units.

For example, you can estimate the mass of a pencil. Put a pencil in one hand and a gram cube in the other hand. Estimate how many gram cubes will equal the mass of the pencil. Measure to see if your estimate was correct.

Practice Estimating and Measuring

Materials

- balance
- gram cubes
- classroom objects

Follow This Procedure

1 Make a chart like the one below.

Object	Estimated mass in grams	Actual mass in grams

2 Hold an object in one hand and gram cubes in the other.

3 Estimate the mass of the object in grams. Add and remove gram cubes from your hand to equal your estimate.

4 Place the object on one side of the balance and place gram cubes on the other side to find out the actual mass of the object in grams.

5 Record your data in the chart.

6 Repeat steps 2 to 5 for each object.

Thinking About Your Thinking

Which object has the most mass? Which object has the least mass? How close were your estimates to the actual masses?

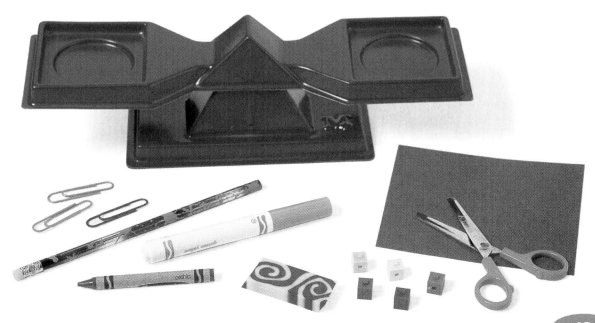

Inferring

How do you infer?

Has anyone ever asked you to make a guess about something? When you make a reasonable guess based on what you have observed or what you know, you are making an inference.

Look at the picture in this page. What is the girl pouring on the plant? How can you tell? What made you guess the way you did?

Practice Inferring

Materials

- *magnet* (labeled)
- metal bolt
- metal nails
- cardboard
- small wooden plank
- paper clips
- paper

Follow This Procedure

1 Take the object labeled *magnet* and test it with the metal objects. Observe if the object labeled *magnet* attracts the metal objects.

2 Infer from your observations and past experience if the object labeled *magnet* is a magnet.

3 Place a piece of paper over the metal objects. Observe if the *magnet* attracts the metal objects. Consider your observations and past experience. What can you infer about the *magnet* and the paper? Does the *magnet* attract objects through the paper?

4 Place a piece of cardboard over the metal objects. Observe if the *magnet* attracts the metal objects. Consider your observations and past experience. What can you infer about the *magnet* and the cardboard? Does the *magnet* attract objects through cardboard?

5 Place a piece of wood over the metal objects. Observe if the *magnet* attracts the metal objects. Consider your observations and past experience. What can you infer about the *magnet* and the wood? Does the *magnet* attract objects through wood?

Thinking About Your Thinking

List the steps you used to make an inference about how magnets attract through paper. Do you think it is always correct?

Can you infer from these observations that magnetic attraction never goes through wood? Just this magnet? Just this wood?

Predicting

How can you make predictions?

When you predict, you make a forecast about what will happen in the future. Predictions are based on what you have studied or observed.

Read each group of sentences below. On your own paper, predict what will happen.

A stalk of celery was put into a glass jar. The jar had red water in it. Think about what you know about plants. Predict what will happen.

It has rained all morning. A puddle is on the sidewalk. The sun comes out and the day is bright and warm. Predict what will happen to the puddle.

Practice Predicting

Materials

- 20 mL of water
- 100 mL graduated cylinder
- 25 marbles

Follow This Procedure

1 Make a chart like the one below.

Number of marbles	Prediction	Water Level in mL
0		
5		
10		
15		
20		
25		

2 Make a graph like the one below.

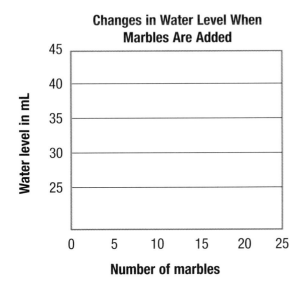

Changes in Water Level When Marbles Are Added

3 Pour 20 mL of water into the graduated cylinder.

4 Record the water level in the chart.

5 Tilt the cylinder and carefully drop 5 marbles into it. Record the water level in the chart.

6 Add 5 more marbles so you have 10 marbles in the cylinder. Again, record the water level.

7 Predict what the water level will be with 15 marbles. Predict what the water level will be with 20 marbles and with 25 marbles.

8 Test each of your predictions and record your results.

9 Fill in your graph with the data from your chart.

Thinking About Your Thinking

What do you think will be the water level of the cylinder if you added 30 marbles? 40 marbles? What did you use to make these predictions?

Making Operational Definitions

What is an operational definition?

An operational definition is a definition or description of an object or an event based on your experience with it.

For example, an operational definition of a clock is: an object that has hands and keeps time in hours.

As your experience changes, your operational definition may also change. For example, you then might give this definition of a clock: a device that keeps time electronically and has digital numbers.

A good operational definition can shed light on even the most difficult topics. Can you define the word "shadow" based on what you know about it?

Practice Making Operational Definitions

Materials

- flashlight
- meter stick or tape measure

Follow This Procedure

1 Write what you know about shadows. Are they light or dark? What causes their shapes? When don't you have a shadow?

2 Early in the morning, go out to the schoolyard as a class and look for your shadow. Have a classmate measure the length of your shadow.

3 Return to the same spot at noon and measure your shadow again. Is it shorter or longer? Is the sun higher or lower at noon than in the morning?

4 Explore making shadows with the flashlight and your fingers. Can you make the shadows longer or shorter? How?

5 Write an operational definition of the word "shadow" based on your experience.

Thinking About Your Thinking

Where is your shadow when there is no light? Did your operational definition help you to answer the question?

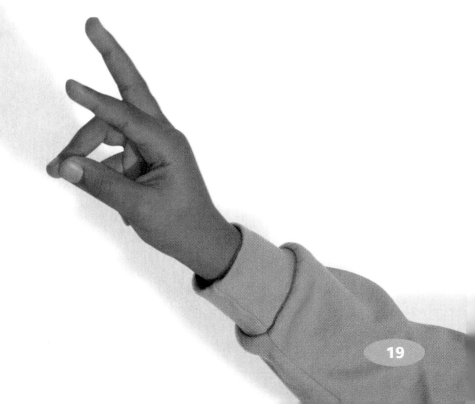

Making and Using Models

How can you make and use models?

A model is a small copy of something. Models are important tools for explaining ideas, objects, and events.

Anything that is not real but is a copy of an actual idea, object, or event can be called a model. There are four steps to making a model.

1. Learn all about an object or event.

2. Think about what you could do to make a copy of the object or event.

3. Make the model.

4. Compare your model to the actual object or event.

Practice Making and Using Models

Materials

- cardboard
- tape
- small rubber or plastic wheels
- paint brushes
- string
- paper
- scissors
- toothpicks
- plastic wrap
- paint
- fabric
- rubber bands

Follow This Procedure

1 Begin by imagining the model car you wish to build. Keep in mind the materials that are available. Draw your model car.

2 Start making the body of your model car. Do you want your car to be very light, or more sturdy? Do you want to be able to see everything inside of it?

3 Will your car have doors? If so, how many? Will there be a roof? Can the roof come off as in a convertible?

4 How will your car be powered? Where is the power source located?

5 Try moving the car without wheels. Will wheels make movement easier or more difficult? Add wheels to your model.

6 What color is your car? Paint your car or cover it with fabric.

Thinking About Your Thinking

How did making a model of a car help you to learn about real cars? What other models have you built?

Formulating Questions and Hypotheses

How can you formulate questions and hypotheses?

Scientific inquiry often begins with asking a question. Sometimes one question leads to an even more useful one. To answer your question, you formulate a hypothesis and design an experiment to test it.

Keep the question clearly in focus as you formulate your hypothesis. Testing a well thought out hypothesis will help you with your scientific inquiry.

Practice Formulating Questions and Hypotheses

Materials

- container of water
- 4 sponges
- construction paper
- graduated cylinder

Follow This Procedure

1 **Question:** How does the size of a sponge affect the amount of water it will hold? Write down your hypothesis or educated guess.

2 Make a chart like the one below.

Sponge size	Water left in cylinder (mL)	Water in sponge (mL)
2 cm x 2 cm		
3 cm x 3 cm		
4 cm x 4 cm		
5 cm x 5 cm		

3 Place the 2 cm × 2 cm sponge on a dry piece of construction paper.

4 Fill the graduated cylinder with water to the 250 mL mark.

5 Pour the water 10 mL at a time evenly over the sponge. Each time, lift the sponge. When the paper is wet, stop pouring. Record the amount of water left in the cylinder.

6 Subtract the amount of water left in the cylinder from 250 to find out how much water is in the sponge. Record this number in the chart.

7 Refill the graduated cylinder to 250 mL.

8 Repeat steps 3 - 7 with the remaining three sponges.

Thinking About Your Thinking

Did your investigation support your hypothesis? Explain.

Collecting and Interpreting Data

How do you collect and interpret data?

You collect data when you observe things and make measurements. The data can be put into graphs, tables, charts, or diagrams. You interpret data when you use the information to solve problems or to answer questions.

Rice cereal

Nutrition Facts		
Serving Size:		11/4 Cup (33g/1.2 oz.)
Servings per Package		About 16
Amount per Serving	Cereal	Cereal with 1/2 Cup Vitamins A & D Skim Milk
Calories	120	160
Fat Calories	0	0
	% Daily Value**	
Total Fat 0g*	0%	0%
Saturated Fat 0g	0%	0%
Cholesterol 0mg	0%	0%
Sodium 350mg	15%	17%
Potassium 40mg	1%	7%
Total Carbohydrates 29g	10%	11%
Dietary Fiber 0g	0%	0%
Sugars 3g		
Other Carbohydrate 25g		
Protein 2g		

Wheat cereal

Nutrition Facts		
Serving Size:		1 Cup (29g/1.0 oz.)
Servings per Package		About 12
Amount per Serving	Cereal	Cereal with 1/2 Cup Vitamins A & D Skim Milk
Calories	110	150
Fat Calories	0	0
	% Daily Value**	
Total Fat 0g*	0%	0%
Saturated Fat 0g	0%	0%
Cholesterol 0mg	0%	0%
Sodium 210mg	9%	11%
Potassium 35mg	1%	7%
Total Carbohydrates 25g	8%	10%
Dietary Fiber 1g	4%	4%
Sugars 3g		
Other Carbohydrate 21g		
Protein 2g		

Corn cereal

Nutrition Facts		
Serving Size:		1 Cup (31g/1.1 oz.)
Servings per Package		About 14
Amount per Serving	Cereal	Cereal with 1/2 Cup Vitamins A & D Skim Milk
Calories	120	160
Fat Calories	0	0
	% Daily Value**	
Total Fat 0g*	0%	0%
Saturated Fat 0g	0%	0%
Cholesterol 0mg	0%	0%
Sodium 120mg	5%	8%
Potassium 25mg	1%	7%
Total Carbohydrates 28g	9%	11%
Dietary Fiber 0g	0%	0%
Sugars 14g		
Other Carbohydrate 14g		
Protein 1g		

Practice Collecting and Interpreting Data

Materials

- graph paper
- pencil

Follow This Procedure

1 Make a chart like the one shown. Look over the Nutrition Facts on each cereal box shown on the page on the left. Record the following data from each of them: amount per serving of sodium (in mg), protein (in g), and sugars (in g).

Amount per serving

Cereal	Sodium (mg)	Protein (g)	Sugars (g)
Wheat			
Rice			
Corn			

2 Based on the information in your chart, which cereals have the same amount of protein? Which cereal has the most sugars?

3 Make a bar graph that compares the amount of sodium in the three cereals. Use the graph below as a model.

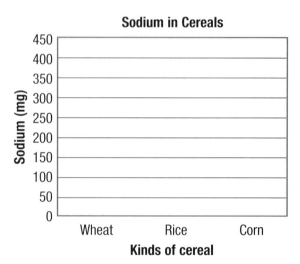

4 From your data, determine which cereal has the most sodium. Which has the least?

Thinking About Your Thinking

How might you use a bar graph to show the amount of potassium in each cereal?

Identifying and Controlling Variables

How do you identify and control variables?

A variable is anything that can change in an experiment. You identify and control variables when you change one variable that may affect the result of an experiment. You control the other variables by keeping them the same.

First you should decide which variable you wish to change. Then you identify all variables that you need to keep the same.

Practice Identifying and Controlling Variables

Materials

- 6 books
- ruler with groove down the middle
- marble
- metric tape measure

Follow This Procedure

1. Make a chart like the one below.

Height of ramp	Distance marble travels (cm)
1 book	
2 books	
3 books	
4 books	
5 books	
6 books	

2. Make a ramp by setting one end of the ruler on top of one of the books. Place the other end of the ruler on a lower, flat surface.

3. Roll the marble from the top of the ramp. Use the metric tape measure to measure the distance the marble travels from the end of the ramp.

4. Record the distance in your chart.

5. Add another book and roll the marble again. Measure the distance it travels from the end of the ramp.

6. Repeat this procedure until the ramp is 6 books high.

Thinking About Your Thinking

Which variable did you change? Which variable responded to the change, or, in other words, what did you measure? Which variables were kept the same?

Experimenting

How do you perform a scientific experiment?

In a scientific experiment, you design an investigation to try to solve a problem by testing a hypothesis. Based on the results, you draw conclusions.

Here are the steps in the process:

1. State the problem you are investigating.

2. Formulate a hypothesis about the problem.

3. Identify and control the variables. Decide which variables you will keep the same and which you will change.

4. Test your hypothesis in an experiment.

5. Collect your data.

6. Interpret your data.

7. State your conclusion. Did the data support your hypothesis?

Practice Experimenting

Materials

- safety goggles
- marble
- paper
- oil
- water
- two 50 mL cylinders
- stopwatch

Follow This Procedure

1 Think about solid objects falling through liquids, like a toy falling through bath water. Would the toy fall as quickly through oil? Which is thicker, oil or water?

2 State the problem. What factors affect the rate at which an object falls through a liquid?

3 Write a hypothesis about the thickness of a liquid and the time it would take for a marble to drop to the bottom of a cylinder filled with that liquid.

4 Design your experiment. The variable that will change is the type of liquid in the cylinder. The marble and the amount of liquid will be the variables that remain the same.

NOTE: *Gently* push the marble from the edge of each cylinder into the liquid. *Don't* throw it with any force.

5 Make a chart to record your data.

6 Put on your safety goggles. Perform your experiment. Record the time it takes for the marble to reach the bottom of the cylinder of each liquid. Do several trials for each liquid.

7 Interpret you data by making a graph based on information in your chart.

8 Make a conclusion about your hypothesis. Does the data support your hypothesis or prove it wrong?

Thinking About Your Thinking

List the seven steps in the logical process of performing a scientific experiment. Is their order important?

Kingdoms of Living Things

Scientists divide the millions of organisms that live on the earth into five groups called kingdoms. Organisms in each kingdom are like each other in some ways and are different from organisms in other kingdoms. The five kingdoms of living things are monerans, protists, fungi, plants, and animals.

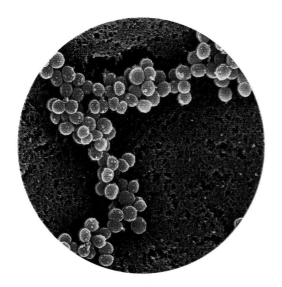

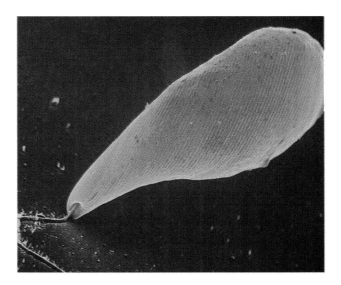

Monerans

▲ *Monerans are made of one cell. They are so small that they cannot be seen without a microscope. Bacteria and other one-celled organisms belong to this kingdom.*

Protists

▲ *Some protists are made of one cell. Others are made of many cells. Many protists have parts that help them move. Certain protists can make their own food while others must take in food. Most seaweeds are protists. Seaweeds are made of many cells and can make their own food.*

Fungi

▲ *Some fungi are one-celled organisms. Most fungi, however, are made of more than one cell. Fungi cannot make their own food. They absorb food made by other organisms. Molds, yeasts, and mushrooms are examples of fungi.*

Animals

▲ *Like plants, animals are made of many cells. Animals cannot make their own food. Most animals can move from place to place.*

Plants

Plants are made of many cells. Green plants use sunlight to make their own food. Plants cannot move from place to place. Their roots hold them tightly in the soil. ▼

Vertebrates and Invertebrates

The animal kingdom can be divided into two main groups. One group contains animals that have a backbone. Animals that have backbones are called vertebrates. The other group contains animals that do not have a backbone. These animals are called invertebrates.

▲ Crabs belong to a group called crustaceans.

Invertebrates

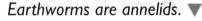
Earthworms are annelids. ▼

Spiders belong to a group of invertebrates called arachnids. ▶

The group to which sponges belong is known as porifera. ▼

▲ Insects are the largest group of animals.

◀ Jellyfish are coelenterates.

Vertebrates

◄ Snakes, turtles, and lizards belong to a group called reptiles.

▲ A hummingbird is one of many different birds.

Koalas are mammals. ▶

Frogs are amphibians. ▶

Sharks are fishes. ▶

Life Cycle of a Tree

The stages in the life cycle of a tree include seeds, germination, seedling, and growth and pollination. After a time, the tree dies, falls to the ground, and begins to rot, or decompose.

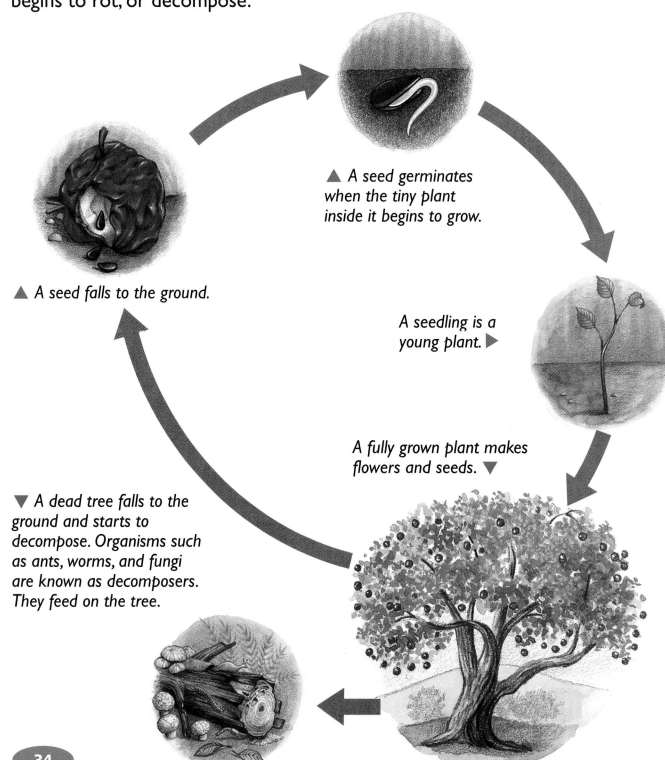

▲ A seed germinates when the tiny plant inside it begins to grow.

▲ A seed falls to the ground.

A seedling is a young plant. ▶

A fully grown plant makes flowers and seeds. ▼

▼ A dead tree falls to the ground and starts to decompose. Organisms such as ants, worms, and fungi are known as decomposers. They feed on the tree.

Mixtures and Solutions

Mixtures

A mixture is formed when two or more objects are mixed together, but each object keeps its own properties. You can easily separate the different parts that make up a mixture.

▲ *A salad is a mixture. Even though all the parts taste good together, each piece has its own flavor.*

◀ *Pasta with sauce is a mixture. The sauce itself is a mixture too.*

Solutions

A solution is a special kind of mixture. In a solution, two or more substances are evenly mixed together. We sometimes say that one material is dissolved in another. Some properties of some of the substances might change.

In soda water, a gas is dissolved in water. You cannot see the gas until it starts to separate from the water. When this happens, you see the gas as bubbles in the water. ▼

Ocean water is a solution that contains many minerals dissolved in water. ▼

Properties of Matter

Some properties of matter can be measured. Other properties cannot be measured but can be described. Color, size, and taste are some properties of matter. Magnetism, hardness, and density are also properties of matter.

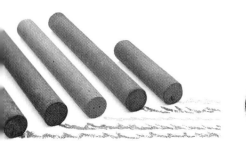

▲ *This rock, called a lodestone, is a mineral that is found in the earth. Lodestones are natural magnets. They will push together or pull apart from each other. Materials made of iron or with iron in them will be attracted or repelled by lodestones.*

▲ *You can tell how hard a material is by rubbing it against another material. The harder material will scratch the softer material. The objects shown here have different levels of hardness. The chalk is the softest, followed by the pencil lead, and then the penny. A diamond is the hardest mineral. It will scratch any other material.* ▼

Density describes how much mass is in a certain amount of matter. The rubber duck and the metal car are about the same size. The density of the car is greater than the density of water. The density of the rubber duck is less than the density of water. Therefore, the rubber duck will float on the surface of the water while the car sinks to the bottom. ▼

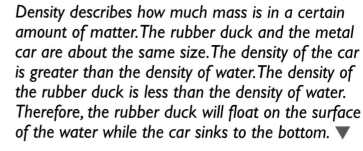

Structure of Matter

All matter is made of smaller and smaller pieces, or particles. You cannot see the smallest particles of matter. An atom is the smallest whole bit of each kind of matter. Two or more atoms can join together to form larger particles. In the different states of matter, the particles are arranged in different ways.

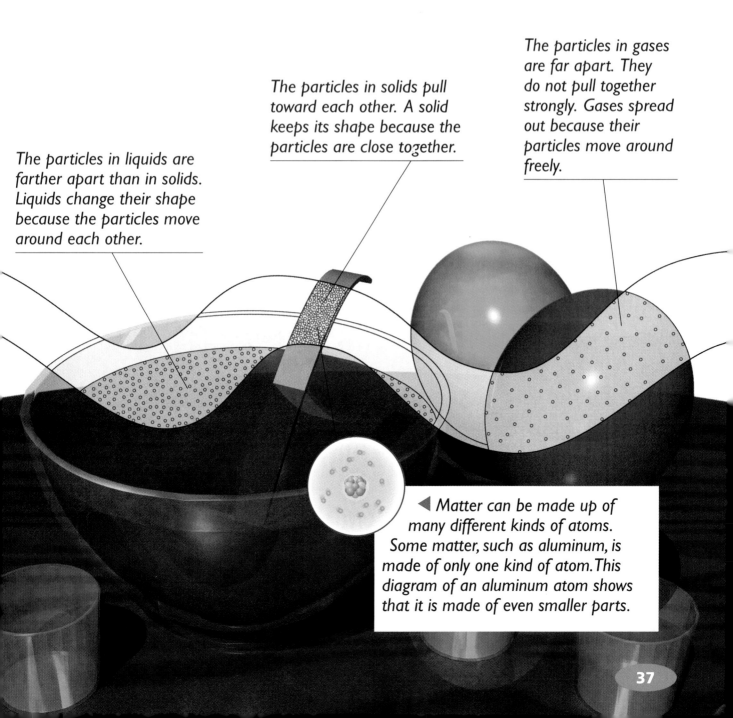

The particles in gases are far apart. They do not pull together strongly. Gases spread out because their particles move around freely.

The particles in solids pull toward each other. A solid keeps its shape because the particles are close together.

The particles in liquids are farther apart than in solids. Liquids change their shape because the particles move around each other.

◀ *Matter can be made up of many different kinds of atoms. Some matter, such as aluminum, is made of only one kind of atom. This diagram of an aluminum atom shows that it is made of even smaller parts.*

Waves of Energy

Waves carry energy from one place to another. Sound and light are forms of energy that move as waves. You can see some waves, such as waves in water. Other waves, such as sound waves, are invisible.

◀ *Sound waves move out from a gong after it has been hit. The sound waves travel from the vibrating gong through the air to your ear.*

When you drop a stone into water, waves of energy travel through the water. The energy waves move out from the center in the form of circles. ▼

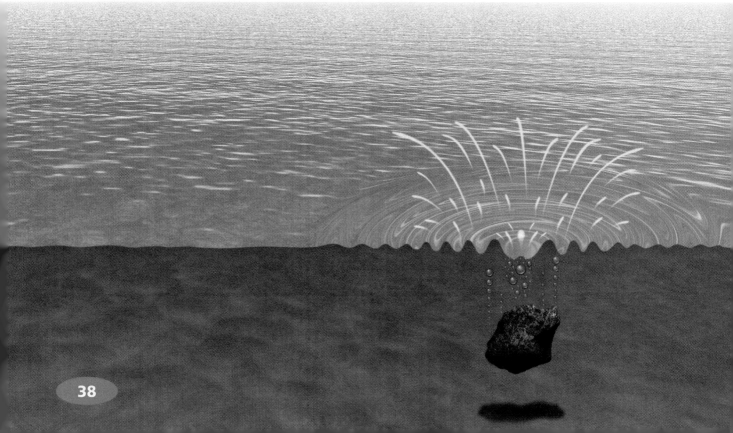

Colors of White Light

The light that we see is known as white light. White light really is made up of many different colors.

Light energy travels in waves. The colors that make up white light can be separated into a band of different colors. This band is called the visible spectrum. The seven colors of the visible spectrum are red, orange, yellow, green, blue, indigo, and violet.

▶ A prism, or piece of glass or plastic shaped like a triangle, can separate white light into the colors of the visible spectrum.

◀ You see a rainbow when raindrops in the air bend the light waves coming from the sun. The sunlight is separated into the different colored parts of the visible spectrum.

Systems

A system is a set of things that form a whole. Systems can be made of many different parts. All the parts depend on each other. The whole system works because all the parts work together.

▼ *A sprouted seed is a simple system. Each part of the sprouted seed depends on other parts. For example, the roots of the sprout get water and minerals from the soil. The stem carries water, minerals, and sugars to the other parts of the sprout. The leaves make sugars.*

◄ *This toy is a simple system that is made up of a piece of wood attached to a piece of string. Each part of the system is needed to make the toy work the way it does.*

◀ This ecosystem is an example of a system that contains living things. The parts of the system that interact and depend on each other are the plants and the buffalo.

◀ The fish, plants, water, and oxygen in this aquarium all depend on each other to make the system work.

Layers of the Earth

Atmosphere

A blanket of air, called the atmosphere, surrounds the earth. The earth's atmosphere protects it from harmful sunlight and helps organisms on the earth survive.

Crust

The earth itself is made of layers. The outer layer, or crust, of the earth is made up of rocks and soil. The land you walk on and the land under the oceans are part of the crust.

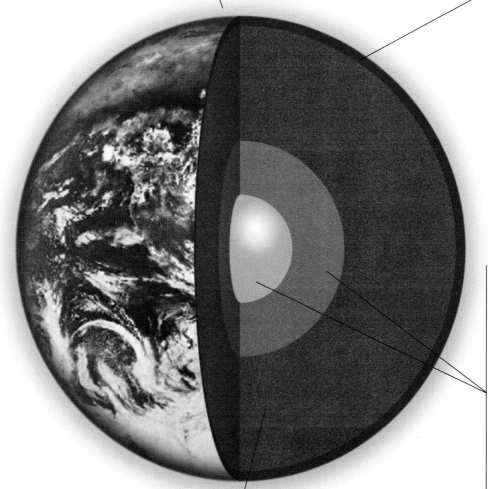

Core

The center of the earth—the core—is made mostly of iron. The outside part of the core has liquid iron. The inside part has solid iron. The core is the hottest part of the earth. The temperature of the core is almost as hot as the surface of the sun!

Mantle

The middle layer of the earth is called the mantle. The mantle is mostly made of rock. Some of the rock in the mantle is partly melted.

The Rock Cycle

In the rock cycle, rocks form and change into other types of rock. Rocks form in three main ways. Over millions of years, each type of rock can change into another type of rock.

Rocks that form from melted material deep inside the earth are igneous rocks. Granite is an igneous rock.

As a result of weathering, rocks break down. Sand and small bits of rock sink beneath the water. Layers of material press together underwater and form sedimentary rocks. Sandstone is a sedimentary rock.

Metamorphic rock forms as very high heat and great pressure within the earth change igneous and sedimentary rocks. Gneiss is a metamorphic rock.

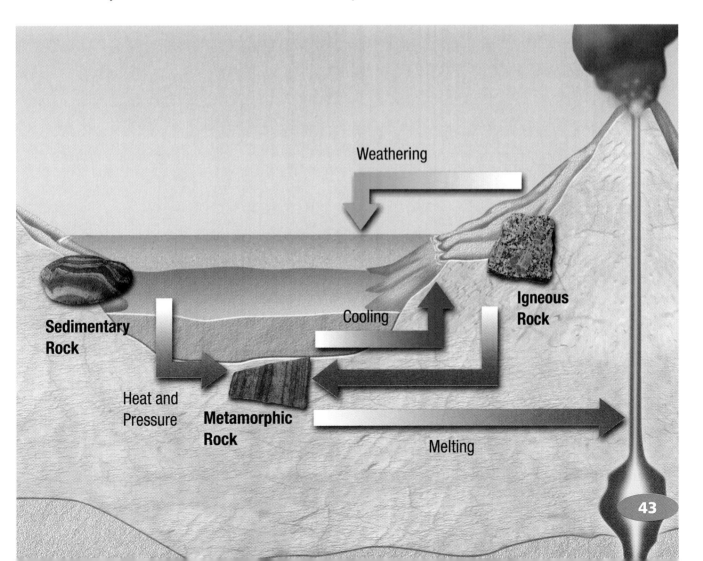

Weathering

Cooling

Igneous Rock

Sedimentary Rock

Heat and Pressure

Metamorphic Rock

Melting

Constellations

Long ago, people divided the stars in the sky into groups. They connected each group of stars with imaginary lines. The stars and lines looked like pictures in the sky. Each group of stars connected by imaginary lines is called a constellation.

The Big Dipper

Ursa Major or the Great Bear ▼

Canis Major ▼

Orion the Hunter ▲

Eclipses of the Sun and Moon

In an eclipse of the sun, or solar eclipse, the moon comes between the sun and Earth. The moon then makes a shadow form on Earth. Sometimes the moon blocks all of the sunlight from certain places on Earth. People in these places see a total solar eclipse. Daytime seems as dark as night. In other places, only some light is blocked. People there see a partial eclipse.

Solar Eclipse

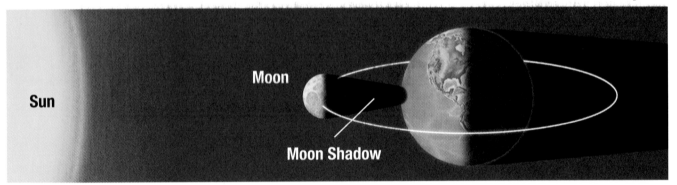

Sun

Moon

Moon Shadow

Sometimes Earth comes between the moon and the sun. When the moon moves through the shadow of Earth, an eclipse of the moon, or a lunar eclipse happens.

Lunar Eclipse

Sun

Moon

Earth Shadow

Natural Resources

Natural resources are useful materials that come from the earth. Natural resources can be renewable, nonrenewable, or inexhaustible.

▲ Trees are a renewable resource. New trees can be planted to replace trees that have been cut down. It may take years, however, for the new trees to become fully grown. The soil that the trees grow in is also a renewable resource. Weathered rock and humus are always being added to soil.

▲ Some natural resources, such as gas and oil, are nonrenewable. They cannot be replaced. Once the supply of gas and oil from the earth is used up, these resources will be gone forever.

▲ Sunlight and water are examples of natural resources that can never be used up. They are known as inexhaustible natural resources.

▶ Wind is also an inexhaustible natural resource. Wind is being used here to make electric power.

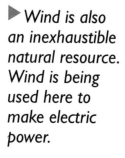

Tools

Tools can make objects appear larger. They can help you measure volume, temperature, length, distance, and mass. Tools can help you figure out amounts and analyze your data. Tools can also provide you with the latest scientific information.

▲ Safety goggles protect your eyes.

You can figure amounts using a calculator. ▶

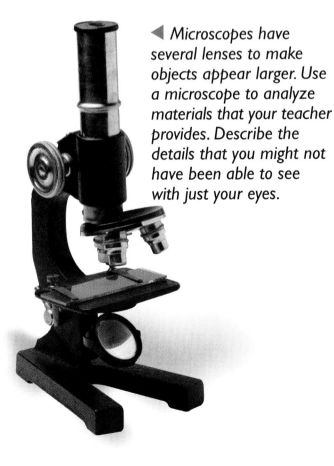

◀ *Microscopes have several lenses to make objects appear larger. Use a microscope to analyze materials that your teacher provides. Describe the details that you might not have been able to see with just your eyes.*

▲ *A hand lens makes objects appear larger so you can see more details.*

▲ Computers can quickly provide the latest scientific information.

▶ You use a thermometer to measure temperature. Many thermometers have both Farenheit and Celsius scales. Usually scientists only use the Celsius scale when measuring temperature.

Scientists use metric rulers and meter sticks to measure length and distance. Scientists use the metric units of meters, centimeters, and millimeters to measure length and distance. ▼

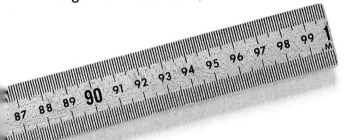

▼ Use a camera to collect and analyze information about a plant. Take a picture of a green plant that is near a sunny window. Do not move or turn the plant. Take another picture of it a week later. Compare the pictures to see if the plant has changed.

Clocks are used for measuring time. ▼

▲ *You can talk into a tape recorder to record information you want to remember. Use a tape recorder to collect and analyze information about the sounds of different kinds of birds. Then see if your classmates can identify the birds.*

▲ *You use a balance to measure mass.*

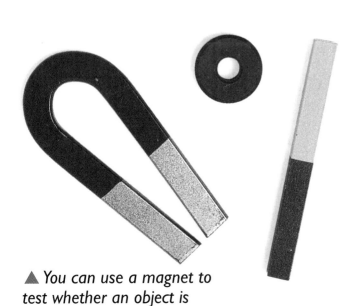

▲ *You can use a magnet to test whether an object is made of certain metals such as iron.*

▲ *A compass is used to indicate direction. The directions on a compass include north, south, east, and west.*

| 8000 B.C. | 6000 B.C. | 4000 B.C | 2000 B.C. |

**Life
Science**

**Physical
Science**

● **3000 B.C.**
The Egyptians develop
geometry. They use it
to re-measure their
farmlands after floods
of the Nile River.

**Earth
Science**

● **8000 B.C.** Farming
communities start as people
use the plow for farming.

**Human
Body**

4th century B.C.
Aristotle classifies
plants and animals.

3rd century B.C.
Aristarchus proposes that the
earth revolves around the sun.

4th century B.C.
Aristotle describes the
motions of falling
bodies. He believes that
heavier things fall faster
than lighter things.

260 B.C. Archimedes
discovers the principles of
buoyancy and the lever.

4th century B.C. Aristotle
describes the motions
of the planets.

200 B.C. Eratosthenes calculates
the size of the earth. His result is
very close to the earth's actual
size.

87 B.C.
Chinese report observing
an object in the sky that
later became known as
Halley's comet.

5th and 4th centuries B.C.
Hippocrates and other Greek
doctors record the symptoms of
many diseases. They also urge
people to eat a well-balanced diet.

Life Science

Physical Science

83 A.D. Chinese travelers use the compass for navigation.

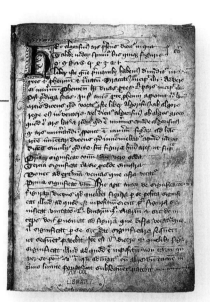

About 750–1250 Islamic scholars get scientific books from Europe. They translate them into Arabic and add more information.

Earth Science

140 Claudius Ptolemy draws a complete picture of an earth-centered universe.

132 The Chinese make the first seismograph, a device that measures the strength of earthquakes.

Human Body

2nd century Galen writes about anatomy and the causes of diseases.

1100s
Animal guide books begin to appear. They describe what animals look like and give facts about them.

1250
Albert the Great describes plants and animals in his book *On Vegetables and On Animals.*

1555
Pierre Belon finds similarities between the skeletons of humans and birds.

9th century
The Chinese invent block printing. By the 11th century, they had movable type.

1019
Abu Arrayhan Muhammad ibn Ahmad al'Biruni observed both a solar and lunar eclipse within a few months of each other.

1543
Nikolaus Copernicus publishes his book *On The Revolutions of the Celestial Orbs.* It says that the sun remains still and the earth moves in a circle around it.

1265
Nasir al-Din al-Tusi gets his own observatory. His ideas about how the planets move will influence Nikolaus Copernicus.

About 1000
Ibn Sina writes an encyclopedia of medical knowledge. For many years, doctors will use this as their main source of medical knowledge. Arab scientist Ibn Al-Haytham gives the first detailed explanation of how we see and how light forms images in our eyes.

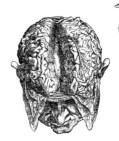

1543
Andreas Vesalius publishes *On the Makeup of the Human Body.* In this book he gives very detailed pictures of human anatomy.

1600	1620	1640	1660	1680

Life Science

1663 Robert Hooke first sees the cells of living organisms through a microscope. Antoni van Leeuwenhoek discovers bacteria with the microscope in 1674.

1679 Maria Sibylla Merian paints the first detailed pictures of a caterpillar turning into a butterfly. She also develops new techniques for printing pictures.

Physical Science

1600 William Gilbert describes the behavior of magnets. He also shows that the attraction of a compass needle toward North is due to the earth's magnetic pole.

1632 Galileo Galilei shows that all objects fall at the same speed. Galileo also shows that all matter has inertia.

1687 Isaac Newton introduces his three laws of motion.

Earth Science

1609–1619 Johannes Kepler introduces the three laws of planetary motion.

1610 Galileo uses a telescope to see the rings around the planet Saturn and the moons of Jupiter.

1669 Nicolaus Steno sets forth the basic principles of how to date rock layers.

1650 Maria Cunitz publishes a new set of tables to help astronomers find the positions of the planets and stars.

1693–1698 Maria Eimmart draws 250 pictures depicting the phases of the moon. She also paints flowers and insects.

1687 Isaac Newton introduces the concept of gravity.

Human Body

1628 William Harvey shows how the heart circulates blood through the blood vessels.

1700	1720	1740	1760	1780	1800

1735 Carolus Linnaeus devises the modern system of naming living things.

1704 Isaac Newton publishes his views on optics. He shows that white light contains many colors.

1759 Emile du Châtelet translates Isaac Newton's work into French. Her work still remains the only French translation.

1789 Antoine-Laurent Lavoisier claims that certain substances, such as oxygen, hydrogen, and nitrogen, cannot be broken down into anything simpler. He calls these substances "elements."

1729 Stephen Gray shows that electricity flows in a straight path from one place to another.

1781 Caroline and William Herschel (sister and brother) discover the planet Uranus.

1784 French chemist Antoine-Laurent Lavoisier does the first extensive study of respiration.

1798 Edward Jenner reports the first successful vaccination for smallpox.

1721 Onesimus introduces to America the African method for inoculation against smallpox.

55

1805	**1810**	**1815**	**1820**	**1825**	**1830**	**1835**

Life Science

1808 French naturalist Georges Cuvier describes some fossilized bones as belonging to a giant, extinct marine lizard.

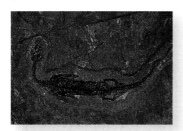

1838–1839 Matthias Schleiden and Theodor Schwann describe the cell as the basic unit of a living organism.

Physical Science

1800 Alessandro Volta makes the first dry cell (battery).

1820 H.C. Oersted discovers that a wire with electric current running through it will deflect a compass needle. This showed that electricity and magnetism were related.

1808 John Dalton proposes that all matter is made of atoms.

Earth Science

1830 Charles Lyell writes *Principles of Geology*. This is the first modern geology textbook.

1803 Luke Howard assigns to clouds the basic names that we still use today— cumulus, stratus, and cirrus.

Human Body

1842 Richard Owen gives the name "dinosaurs" to the extinct giant lizards.

1859 Charles Darwin proposes the theory of evolution by natural selection.

1863 Gregor Mendel shows that certain traits in peas are passed to succeeding generations in a regular fashion. He outlines the methods of heredity.

1847 Hermann Helmholtz states the law of conservation of energy. This law holds that energy cannot be created or destroyed. Energy only can be changed from one form to another.

1842 Christian Doppler explains why a car, train, plane, or any quickly moving object sounds higher pitched as it approaches and lower pitched as it moves away.

1866 Ernst Haeckel proposes the term "ecology" for the study of the environment.

Early 1860s Louis Pasteur realizes that tiny organisms cause wine and milk to turn sour. He shows that heating the liquids kills these germs. This process is called pasteurization.

1840s Doctors use anesthetic drugs to put their patients to sleep.

1850s and 1860s Ignaz P. Semmelweis and Sir Joseph Lister pioneer the use of antiseptics in medicine.

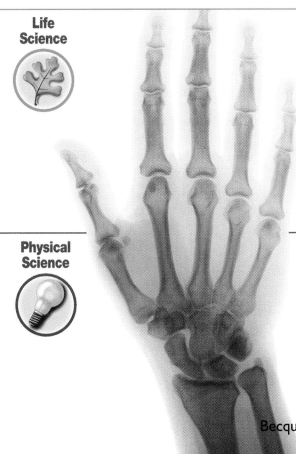

Life Science

● **1900–1910** George Washington Carver, the son of slave parents, develops many new uses for old crops. He finds a way to make soybeans into rubber, cotton into road-paving material, and peanuts into paper.

Physical Science

● **1897** J. J. Thomson discovers the electron.

● **1905** Albert Einstein introduces the theory of relativity.

● **1895** Wilhelm Roentgen discovers X rays.

● **1896** Henri Becquerel discovers radioactivity.

Earth Science

● **1907** Bertram Boltwood introduces the idea of "radioactive" dating. This allows geologists to accurately measure the age of a fossil.

● **1912** Alfred Wegener proposes the theory of continental drift. This theory says that all land on the earth was once a single mass. It eventually broke apart and the continents slowly drifted away from each other.

Human Body

● **1885** Louis Pasteur gives the first vaccination for rabies. Pasteur thought that tiny organisms caused most diseases.

1920s Ernest Everett Just performs important research into how cells metabolize food.

1947 Archaeologist Mary Leakey unearths the skull of a *Proconsul africanus,* an example of a fossilized ape.

1913 Danish physicist Niels Bohr presents the modern theory of the atom.

1911 Ernst Rutherford discovers that atoms have a nucleus, or center.

1911 Marie Curie wins the Nobel Prize for chemistry. This makes her the first person ever to win two Nobel Prizes, the highest award a scientist can receive.

1938 Otto Hahn and Fritz Straussman split the uranium atom. This marks the beginning of the nuclear age.

1942 Enrico Fermi and Leo Szilard produce the first nuclear chain reaction.

1945 The first atomic bomb is exploded in the desert at Alamogordo, New Mexico.

1938 Lise Meitner and Otto Frisch explain how an atom can split in two.

1946 Vincent Schaefer and Irving Langmuir use dry ice to produce the first artificial rain.

1933 Meteorologist Tor Bergeron explains how raindrops form in clouds.

1917 Florence Sabin becomes the first woman professor at an American medical college.

1928 Alexander Fleming notices that the molds in his petri dish produced a substance, later called an antibiotic, that killed bacteria. He calls this substance penicillin.

1935 Chemist Percy Julian develops physostigmine, a drug used to fight the eye disease glaucoma.

1922 Doctors inject the first diabetes patient with insulin.

1950	1955	1960	1965	1970

Life Science

1951 Barbara McClintock discovers that genes can move to different places on a chromosome.

1953 The collective work of James D. Watson, Francis Crick, Maurice Wilkins, and Rosalind Franklin leads to the discovery of the structure of the DNA molecule.

1972 Researchers find human DNA to be 99% similar to that of chimpanzees.

Physical Science

1969 UCLA is host to the first computer node of ARPANET, the forerunner of the internet.

1974 Opening of TRIUMF, the world's largest particle accelerator, at the University of British Columbia.

Earth Science

1957 The first human-made object goes into orbit when the Soviet Union launches *Sputnik I*.

1972 Cygnus X-1 is first identified as a blackhole.

1969 Neil Armstrong is the first person to walk on the moon.

1967 Geophysicists introduce the theory of plate tectonics.

1962 John Glenn is the first American to orbit the earth.

Human Body

1954–1962 In 1954, Jonas Salk introduced the first vaccine for polio. In 1962, most doctors and hospitals substituted Albert Sabin's orally administered vaccine.

1967 Dr. Christiaan Barnard performs the first successful human heart transplant operation.

1964 The surgeon general's report on the hazards of smoking is released.

NO SMOKING
American Cancer Society

1975   **1980**   **1985**   **1990**   **1995**   **2000**

1988
Congress approves funding for the Human Genome Project. This project will map and sequence the human genetic code.

1997
Scientists in Edinburgh, Scotland, successfully clone a sheep, Dolly.

1975 People are able to buy the first personal computer, called the Altair.

1996 Scientists make "element 112" in the laboratory. This is the heaviest element yet created.

1979 A near meltdown occurs at the Three Mile Island nuclear power plant in Pennsylvania. This alerts the nation to the dangers of nuclear power.

Early 1990s The first "extra-solar" planet is discovered.

1995 The National Severe Storms Laboratory develops NEXRAD, the national network of Doppler weather radar stations for early severe storm warnings.

1976 National Academy of Sciences reports on the dangers of chlorofluorocarbons (CFCs) for the earth's ozone layer.

1981 The first commercial Magnetic Resonance Imaging scanners are available. Doctors use MRI scanners to look at the non-bony parts of the body.

1982 Dr. Stanley Prusiner identifies a new kind of disease-causing agent—prions. Prions are responsible for many brain disorders.

1998 John Glenn, age 77, orbits the earth aboard the space shuttle *Discovery*. Glenn is the oldest person to fly in space.

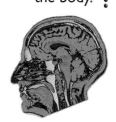

Glossary

Full Pronunciation Key

The pronunciation of each word is shown just after the word, in this way: **ab•bre•vi•ate** (ə brē′vē āt).

The letters and signs used are pronounced as in the words below.

The mark ′ is placed after a syllable with primary or heavy accent, as in the example above.

The mark ′ after a syllable shows a secondary or lighter accent, as in **ab•bre•vi•a•tion** (ə brē′vē ā′shən).

a	hat, cap	g	go, bag	ō	open, go	ᴛʜ	then, smooth	zh	measure, seizure
ā	age, face	h	he, how	ȯ	all, caught	u	cup, butter		
â	care, fair	i	it, pin	ô	order	u̇	full, put	ə	represents:
ä	father, far	ī	ice, five	oi	oil, voice	ü	rule, move		a in about
b	bad, rob	j	jam, enjoy	ou	house, out	v	very, save		e in taken
ch	child, much	k	kind, seek	p	paper, cup	w	will, woman		i in pencil
d	did, red	l	land, coal	r	run, try				o in lemon
e	let, best	m	me, am	s	say, yes	y	young, yet		u in circus
ē	equal, be	n	no, in	sh	she, rush	z	zero, breeze		
ėr	term, learn	ng	long, bring	t	tell, it				
f	fat, if	o	hot, rock	th	thin, both				

A

adaptation (ad′ap tā′shən), a structure or behavior that helps an organism survive in its environment.

alcohol (al′kə hȯl), a drug found in beer, wine, and liquor that can be harmful.

amphibian (am fib′ē ən), an animal with a backbone that lives part of its life cycle in water and part on land.

astronaut (as′trə nȯt), a person who travels in space.

atmosphere (at′mə sfir), air that surrounds the earth.

atom (at′əm), a small particle that makes up matter.

axis (ak′sis), an imaginary straight line through the center of Earth around which Earth rotates.

B

bacteria (bak tir′ē ə), organisms made of one cell that can be seen through a microscope.

bar graph, a graph that uses bars to show data.

blizzard (bliz′ərd), a snowstorm with strong, cold winds and very low temperatures.

C

caption (kap′shən), written material that helps explain a picture or diagram.

carbon dioxide (kär′bən dī ok′sīd), a gas in the air that plants use to make food.

cartilage (kär′tl ij), a tough, rubbery tissue that makes up parts of the skeleton.

cause (kȯz), a person, thing, or event that makes something happen.

cell (sel), the basic unit of all living things, including the human body.

chemical (kem′ə kəl) **change**, a change that causes one kind of matter to become a different kind of matter.

circuit (sėr′kit), the path an electric current follows.

clay (klā) **soil**, soil with tiny grains that are packed closely together.

cloud, a mass of many water droplets or bits of ice that float in the air.

community (kə myü′nə tē), all the plants, animals, and other organisms that live and interact in the same place.

compare (kəm pâr′), to decide which of two numbers is greater.

conclusion (kən klü′zhən), a decision or opinion based on evidence and reasoning.

condense (kən dens′), to change from a gas to a liquid.

conductor (kən duk′tər), a material through which energy flows easily.

conifer (kon′ə fər), a tree or shrub that has cones.

conserve (kən sėrv′), to keep something from being used up.

consumer (kən sü′mər), an organism that eats food.

control (kən trōl′), the part of an experiment that does not have the variable being tested.

core (kôr), the center part of the earth.

crater (krā′tər), a large hole in the ground that is shaped like a bowl.

crust (krust), the solid outside part of the earth.

D

data (dā′tə), information.

decay (di kā′), to slowly break down or rot.

decomposer (dē′kəm pō′zər), a living thing that feeds on the wastes or dead bodies of other living things and breaks them down.

disease (də zēz′), an illness.

drought (drout), a long period of dry weather.

E

eardrum (ir′drum′), the thin, skinlike layer that covers the middle part of the ear and vibrates when sound reaches it.

earthquake (ėrth′kwāk′), a shaking or sliding of the surface of the earth.

echo (ek′ō), a sound that bounces back from an object.

effect (ə fekt′), whatever is produced by a cause; a result.

electric charges (i lek′trik chär′jəz), tiny amounts of electricity present in all matter.

electric circuit (i lek′trik sėr′kit), the path along which electric current moves.

electric current (i lek′trik kėr′ənt), the flow of electric charges.

electromagnet (i lek′trō mag′nit), a metal that becomes a magnet when electricity passes through wire wrapped around it.

embryo (em′brē ō), a developing animal before it is born or hatched.

endangered (en dān′jərd) **organism**, a kind of living thing of which very few exist and that someday might not be found on the earth.

energy (en′ər jē), the ability to do work.

energy of motion (mō′shən), energy that moving objects have.

environment (en vī′rən mənt), all the things that surround an organism.

erosion (i rō′zhən), the carrying away of weathered rocks or soils by water, wind, or other causes.

erupt (i rupt′), to burst out.

evaporates (i vap′ə rāts), changes from a liquid state to a gas state.

extinct (ek stingkt′) **organism**, a kind of living thing that is no longer found on the earth.

F

food chain, the way food passes from one organism to another.

force (fôrs), a push or a pull.

fossil (fos′əl), the hardened parts or marks left by an animal or plant that lived long ago.

friction (frik′shən), the force caused by two objects rubbing together that slows down or stops moving objects.

fuel (fyü′əl), a material that is burned to produce useful heat.

fulcrum (ful′krəm), the point on which a lever is supported and moves.

fungus (fung′gəs), an organism, such as a mold or mushroom, that gets food from dead material or by growing on food or a living thing.

G

gas, a state in which matter has no definite shape or volume.

gear (gir), a wheel with jagged edges like teeth.

germ (jėrm), a thing too tiny to be seen without a microscope; some germs may cause disease.

germinate (jėr′mə nāt), to begin to grow and develop.

gills (gilz), the parts of fish and tadpoles that are used to take in oxygen from water.

glacier (glā′shər), a huge amount of moving ice.

graphic sources (graf′ik sôrs′əz), pictures or diagrams that give information.

gravity (grav′ə tē), the force that pulls objects toward the center of the earth.

H

habitat (hab′ə tat), the place where an organism lives.

humus (hyü′məs), decayed organisms in soil.

hurricane (hėr′ə kān), a huge storm that forms over warm ocean water, with strong winds and heavy rains.

I

illegal (i lē′gəl) **drug**, a drug that is against the law to buy, sell, or use.

inclined plane (in klīnd′ plān), a simple machine that is a flat surface with one end higher than the other.

instinct (in′stingkt), an action that an animal can do without learning.

insulator (in′sə lā′tər), a material through which energy cannot flow easily.

involuntary (in vol′ən ter′ē) **muscle**, the kind of muscle that works without a person's control.

J

joint, the place where two bones come together.

L

landfill (land′fil′), a place where garbage is buried in soil.

landform (land′fôrm′), a shape on the earth's surface.

larva (lär′və), a young animal that has a different shape than the adult.

lava (lä′və), hot, melted rock that comes out of a volcano.

lens (lenz), a piece of material that bends light rays that pass through it.

lever (lev′ər), a simple machine made of a bar or board that is supported underneath at the fulcrum.

life cycle (sī′kəl), the stages in the life of a living thing.

ligament (lig′ə mənt), a strong, flexible tissue that holds bones together at a joint.

liquid (lik′wid), a state in which matter has a definite volume but no shape of its own.

liter (lē′tər), a metric unit of volume or capacity equal to 1,000 mL.

loam (lōm), good planting soil that is a mixture of clay, silt, sand, and humus.

M

magma (mag′mə), hot, melted rock and gases deep inside the earth.

magnet (mag′nit), anything that pulls certain metals, such as iron, to it.

magnetism (mag′nə tiz′əm), the force that causes magnets to pull on objects that are made of certain metals, such as iron.

mammal (mam′əl), an animal with a backbone and hair or fur; mothers produce milk for their babies.

mass, the measure of how much matter an object contains.

matter, anything that takes up space and has weight.

milliliter (mil′ə lē′tər), a metric unit of volume or capacity smaller than a liter.

mineral (min′ər əl), a nonliving material that can be found in soil.

mixture (miks′chər), two or more kinds of matter that are placed together but can be easily separated.

muscle (mus′əl), a body tissue that can tighten or loosen to move body parts.

N

natural resource (nach′ər′əl rē′sôrs), a material that comes from the earth and can be used by living things.

nerve (nėrv), a part of the body that carries messages to the brain.

nicotine (nik′ə tēn′), a drug in tobacco that can harm the body.

nutrient (nü′trē ənt), a mineral that plants and animals need to live and grow; a substance in food that living things need for health and growth.

nymph (nimf), a stage in an insect life cycle between egg and adult that looks like an adult but has no wings.

O

orbit (ôr′bit), the path an object follows as it moves around another object.

ore (or), rock that has a large amount of useful minerals.

organ (ôr′gən), a body part that does a special job within a body system.

organism (ôr′gə niz′əm), a living thing.

over-the-counter medicine (med′ə sən), a medicine that can be bought without a doctor's order.

oxygen (ok′sə jən), a gas in air that living things need to stay alive.

P

petal (pet′l), an outside part of a flower that is often colored.

phase (fāz), the shape of the lighted part of the moon.

physical (fiz′ə kəl) **change**, a change in the way matter looks, but the kind of matter remains the same.

pictograph (pik′tə graf), a graph that uses pictures or symbols to show data.

pitch (pich), how high or low a sound is.

plain (plān), a large, flat area of land.

planet (plan′it), a large body of matter that moves around a star such as the sun.

plateau (pla tō′), a large flat area of land that is high.

pole (pōl), a place on a magnet where magnetism is strongest.

pollen (pol′ən), a fine, yellowish powder in a flower.

pollinate (pol′ə nāt), to carry pollen to the stemlike part of the flower.

pollution (pə lü′shən), anything harmful added to the air, water, or land.

population (pop′yə lā′shən), organisms of the same kind that live in the same place at the same time.

precipitation (pri sip′ə tā′shən), a form of water that falls to the ground from clouds.

predator (pred′ə tər), an organism that captures and eats other organisms.

prediction (pri dik′shən), an idea about what will happen based on evidence.

prescription medicine (pri skrip′shən med′ə sən), a medicine that can be bought only with a doctor's order.

prey (prā), an organism that is captured and eaten by another organism.

producer (prə dü′sər), an organism that makes its own food.

property (prop′ər tē), something about an object—such as size, shape, color, or smell—that you can observe with one or more of your senses.

pulley (pül′ē), a simple machine made of a wheel and a rope.

pupa (pyü′pə), the stage in the insect life cycle between larva and adult.

R

ray, a thin line of light.

recycle (rē sī′kəl), to change something so it can be used again.

reflect (ri flekt′), to bounce back.

revolution (rev′ə lü′shən), movement of an object in an orbit around another object.

rotate (rō′tāt), to spin on an axis.

S

sandy (san′dē) **soil,** loose soil with large grains.

satellite (sat′l īt), an object that revolves around another object.

scale, the numbers that show the units used on a bar graph.

screw (skrü), a simple machine used to hold objects together.

seed coat (sēd kōt), the outside covering of a seed.

seed leaf (sēd lēf), the part inside each seed that contains stored food.

seedling (sēd′ling), a young plant that grows from a seed.

sequence (sē′kwens), one thing happening after another.

simple machine (sim′pəl mə shēn′), one of six kinds of tools with few or no moving parts that make work easier.

solar system (sō′lər sis′təm), the sun, the planets and their moons, and other objects that move around the sun.

solid (sol′id), a state in which matter has a definite shape and volume.

star (stär), a very large mass of hot, glowing gases,

states of matter, the three forms of matter—solid, liquid, and gas.

stored (stôrd) **energy,** energy that can change later into a form that can do work.

system (sis′təm), a group of body parts that work together to perform a job.

T

tadpole (tad′pōl), a very young frog or toad.

telescope (tel′ə skōp), an instrument for making distant objects appear nearer and larger.

temperature (tem′per ə chər), a measure of how hot a place or object is.

tendon (ten′dən), a strong cord of tissue that attaches a muscle to a bone.

tide (tīd), the rise and fall of the ocean, mainly due to the moon's gravity.

tissue (tish′ü), a group of cells that look alike and work together to do a certain job.

tornado (tôr nā′dō), a funnel cloud that has very strong winds and moves along a narrow path.

V

vaccine (vak sēn′), a medicine that can prevent the disease caused by one kind of germ.

vibrate (vī′brāt), move quickly back and forth.

vocal cords (vō′kəl kôrdz), two small folds of elastic tissue at the top of the windpipe.

volcano (vol kā′nō), a type of mountain that has an opening at the top through which lava, ash, or other types of volcanic rock flows.

volume (vol′yəm), the amount of space an object takes up; the loudness or softness of a sound.

voluntary muscle (vol′ən ter′ē mus′əl), the kind of muscle that a person can control.

W

water cycle (sī′kəl), movement of water from the earth to the air and back to the earth.

water vapor (vā′pər), water that is in the form of a gas.

weathering (weTH′ər ing), the breaking apart and changing of rocks.

wedge (wej), a simple machine used to cut or split an object.

wheel and axle (wēl and ak′səl), a simple machine that has a center rod attached to a wheel.

work (wėrk), something done whenever a force moves an object through a distance.

Index

Acknowledgments

Illustration

Borders Patti Green
Icons Precison Graphics

Unit A
10, 14, 22, 23, 24, 25, 27, 38, 39, 63, 71, Precision Graphics
20, 21, 34a, 34b, 34c, 34d, 34f J.B. Woolsey
37, 46, 47 John Edwards
104 John Massie

Unit B
36, 3, Mitchell Heinze
48, 73, 103 J.B. Woolsey
76, 77 Jared Schneidman

Unit C
9a, 62, 66, 67, 68, 70, 75, 89, 100 J.B. Woolsey
105 John Edwards

Unit D
12b, 12c, 13, 17, 21, 31, 39 J.B. Woolsey
15, 19, 20c, 24, 25, 26, 27 John Edwards
36 Robert Lawson
40, 41, 43, 46 William Graham
44 Robert Lawson

Photography

Unless otherwise credited, all photographs are the property of Scott Foresman, a division of Pearson Education.

Cover: Kaku Kurita/Liaison Agency
i Bob Daemmrich Photography
iv T Digital Stock
iv B Zefa Germany/Stock Market
v B Marty Snyderman
vii TR PhotoDisc, Inc.
viii B Alan L. Detrick/Photo Researchers
ix T Robert Frerck/Odyssey Productions
x T K. Ballard/Sharpshooters
x B Mark E. Gibson
xi L Manfred Kage/Peter Arnold, Inc.

Unit A
1 B Ron Sanford/Tony Stone Images
2 T Vincent O'Bryne/Panoramic Images
2 C Lawrence Migdale/Photo Researchers
2 B Wildlife Conservation Society
3 L Wildlife Conservation Society
3 R Living Technologies
15 T Doug Wechsler/Animals Animals/Earth Scenes
15 C J.C. Carton/Bruce Coleman Inc.
16 L Zefa Germany/Stock Market
16 R Dwight Kuhn
17 T Dr. William M. Harlow/Photo Researchers
17 B Dr. William M. Harlow/Photo Researchers
21 Harry Taylor/Oxford Scientific Films/Animals Animals/Earth Scenes
32 Frans Lanting/Minden Pictures
33 T Ed Degginger/Bruce Coleman Inc.
33 C Patrice Ceisel/Stock Boston
33 B Frans Lanting/Minden Pictures
35 BR Runk/Schoenberger/Grant Heilman Photography
35 TR John Colwell/Grant Heilman Photography
36 T E. S. Ross
36 B E. S. Ross
42 Gerald Allen/ENP Images
43 TL Breck P. Kent/Animals Animals/Earth Scenes
43 TR Breck P. Kent/Animals Animals/Earth Scenes
43 BR Zig Leszczynski/Animals Animals/Earth Scenes
43 BL Zig Leszczynski/Animals Animals/Earth Scenes
44 Daniel J. Cox & Associates
45 BR Ed Degginger/Animals Animals/Earth Scenes

45 CR Shin Yoshino/Minden Pictures
45 BL PhotoDisc, Inc.
45 T A. Limbrunner/Bruce Coleman Inc.
45 CL IFA/Bruce Coleman Inc.
48 Ron Kimball
49 TL Frans Lanting/Minden Pictures
49 TR Patti Murray/Animals Animals/Earth Scenes
49 CR Digital Stock
49 B Ralph Reinhold/Animals Animals/Earth Scenes
50 T Superstock, Inc.
50 B Jim Brandenburg/Minden Pictures
51 Steve Robinson/NHPA
53 Daniel J. Cox & Associates
54 T Don Enger/Animals Animals/Earth Scenes
57 Tom Walker/Tony Stone Images
58 Jorgen Vogt(1997)/Image Bank
59 T Superstock, Inc.
59 B Randall B. Henne/Dembinsky Photo Assoc. Inc.
60 TL Frans Lanting/Minden Pictures
60 TR Frans Lanting/Minden Pictures
60 BR Flip Nicklin/Minden Pictures
60 BL Chris Huss/Wildlife Collection
61 TL Rod Planck/TOM STACK & ASSOCIATES
61 TR R. Andrew Odum/Peter Arnold, Inc.
61 C George K. Bryce/Animals Animals/Earth Scenes
61 B PhotoDisc, Inc.
62 T Pat & Tom Leeson/Photo Researchers
62 B Roy Morsch/Bruce Coleman Inc.
63 Leonard Lee Rue III/Animals Animals/Earth Scenes
66 PhotoDisc, Inc.
67 Marty Snyderman
68 T Johnny Johnson/Animals Animals/Earth Scenes
68 T Charles Melton/Wildlife Collection
69 L Rick Poley/Visuals Unlimited
69 R PhotoDisc, Inc.
70 Gerard Lacz/Animals Animals/Earth Scenes
72 R J. C. Carton/Bruce Coleman Inc.
73 L Joe Mcdonald/Bruce Coleman Inc.
73 R Don Skillman/Animals Animals/Earth Scenes
74 R John A. Anderson/Animals Animals/Earth Scenes
74 L Suzanne Murphy/Tony Stone Images
75 Peter Johnson/Corbis Media
76 T David C. Fritts/Animals Animals/Earth Scenes
76 B Fred Whitehead/Animals Animals/Earth Scenes
81 Don Skillman/Animals Animals/Earth Scenes
82 R Martin Miller/Visuals Unlimited
82 L Photography by Thane/Animals Animals/Earth Scenes
86 PhotoDisc, Inc.
87 T Erwin & Peggy Bauer/Bruce Coleman Inc.
87 B Photography by Thane/Animals Animals/Earth Scenes
88 T Richard Packwood/OSF/Animals Animals/Earth Scenes
88 B Charlie Palek/Animals Animals/Earth Scenes
89 Corbis Media
90 PhotoDisc, Inc.
90 Jane Burton/Bruce Coleman Inc.
91 Martin Miller/Visuals Unlimited
92 T Francois Gohier/Photo Researchers
92 BR Lynn Stone/Animals Animals/Earth Scenes
92 BL Tim Davis/Photo Researchers
93 B Illinois State Museum, Springfield
93 TR Granger Collection
93 TL Field Museum of Natural History, Neg. CK30T
94 B. Strickland/NatureScapes/Visuals Unlimited
95 D. Cavagnaro/DRK Photo
96 Joel Sartore/Grant Heilman Photography
97 R Christer Fredriksson/Natural Selection Stock Photography, Inc.

97 L Joseph Sohm/ChromoSohn Inc./Corbis Media
98 R Picture Network International
98 L Bob Daemmrich/Image Works
98 B PhotoDisc, Inc.
99 B Ed Reschke/Peter Arnold, Inc.
99 T Brookfield Zoo/Chicago Zoological Society
102 Robert E. Daemmrich/Tony Stone Images
103 B Chamberlain, M.C./DRK Photo
103 T Donald Specker/Animals Animals/Earth Scenes
104 T PhotoDisc, Inc.
104 B PhotoDisc, Inc.
107 Lynn Stone/Animals Animals/Earth Scenes
112 Public Domain

Unit B
1 Comstock Inc.
2 T Vincent O'Bryne/Panoramic Images
2 C Bill Stormont
2 B Stephen Ferry/Liaison Agency
3 C Phil Degginger/Tony Stone Images
3 B Tom Pidgeon/AP/Wide World
18 Chip Henderson/Tony Stone Images
23 T Superstock, Inc.
29 Chip Henderson/Tony Stone Images
39 Kevin Horan
48 Mike Mazzaschi/Stock Boston
50 R O.S.F./Animals Animals/Earth Scenes
51 T Digital Stock
51 C Johnny Johnson/Animals Animals/Earth Scenes
61 Jeff Greenberg/Visuals Unlimited
62 L Superstock, Inc.
66 TL © Don & Pat Valenti Photography
72 T Michael Newman/PhotoEdit
78 R Photri, Inc.
79 George Diebold/Stock Market
89 T PhotoDisc, Inc.
89 B PhotoDisc, Inc.
90 B Tony Freeman/PhotoEdit
90 T Superstock, Inc.
91 L Shin Yoshino/Minden Pictures
95 T Francois Gohier/Photo Researchers
96 T Bob Kramer/Stock Boston
96 B Bob Kramer/Stock Boston
97 Joe McDonald/Bruce Coleman Inc.
101 B Laura Dwight
104 Breck P. Kent/Animals Animals/Earth Scenes
105 TR C.W. Perkins/Animals Animals/Earth Scenes
105 B Gerard Lacz/Animals Animals/Earth Scenes
105 TL PhotoDisc, Inc.
107 PhotoDisc, Inc.

Unit C
1 T. Snow/Visuals Unlimited
2 T Vincent O'Bryne/Panoramic Images
2 B SIU/Visuals Unlimited
2 C-inset Raphael Gaillarde/Liaison Agency
2 C Arie deZanger for Scott Foresman
3 B National Center for Atmospheric Research
3 C NASA
3 Panoramic Images
8 Jim Sugar/Corbis Media
10 Ragnar Larusson/Photo Researchers
11 Corbis Media
12 Krafft/Explorer/Photo Researchers
14 R Richard Thom/Visuals Unlimited
14 L Andras Dancs/Tony Stone Images
15 C John Warden/Tony Stone Images
15 B Henry W. Robison/Visuals Unlimited
15 T Helberg/Unicorn Stock Photos
22 T Bruce Coleman Inc.
22 B John Lemker/Animals Animals/Earth Scenes
23 T PhotoDisc, Inc.
23 B McCutcheon Alaska U.S.A./Visuals Unlimited